D0948944

ARTIFICIAL YOU

ARTIFICIAL YOU

AI AND THE FUTURE OF YOUR MIND

SUSAN SCHNEIDER

PRINCETON UNIVERSITY PRESS

PRINCETON AND OXFORD

Requests for permission to reproduce material from this work
should be sent to permissions@press.princeton.edu

Published by Princeton University Press
41 William Street, Princeton, New Jersey 08540
6 Oxford Street, Woodstock, Oxfordshire OX20 1TR

press.princeton.edu

ISBN 9780691180144
ISBN (ebook): 9780691197777

British Library Cataloging-in-Publication Data is available

Editorial: Matt Rohal
Production Editorial: Terri O'Prey
Text Design: Leslie Flis
Production: Merli Guerra
Publicity: Sara Henning-Stout, Katie Lewis
Copyeditor: Cyd Westmoreland

Jacket design by Faceout Studio, Spencer Fuller
Jacket background: Wizemark / Stocksy

This book has been composed in Arno Pro and Trade Gothic LT Std

Printed on acid-free paper. ∞

Printed in the United States of America

10 9 8 7 6 5 4 3 2 1

FOR ELENA, ALEX, AND ALLY

CONTENTS

ARTIFICIAL YOU

INTRODUCTION

It is 2045. Today, you are out shopping. Your first stop is the Center for Mind Design. As you walk in, a large menu stands before you. It lists brain enhancements with funky names. "Hive Mind" is a brain chip allowing you to experience the innermost thoughts of your loved ones. "Zen Garden" is a microchip for Zen master-level meditative states. "Human Calculator" gives you savant-level mathematical abilities. What would you select, if anything? Enhanced attention? Mozart-level musical abilities? You can order a single enhancement, or a bundle of several.

Later, you visit the android shop. It is time to buy that new android to take care of the house. The menu of AI minds is vast and varied. Some AIs have heightened perceptual skills or senses we humans lack, others have databases that span the entire Internet. You carefully select the options that suit your family. Today is a day of mind design decisions.

This book concerns the future of the mind. It's about how our understanding of ourselves, our minds, and our nature can drastically change the future, for better or for worse. Our brains evolved for specific environments and are greatly constrained by anatomy and evolution. But artificial intelligence (AI) has opened up a vast design space, offering new materials and modes of operation, as well as novel ways to explore the space at a rate much faster than biological evolution. I call this exciting new enterprise *mind design*. Mind design is a form of intelligent design, but we humans, not God, are the designers.

I find the prospect of mind design humbling, because frankly, we are not terribly evolved. As the alien in the Carl Sagan film *Contact* says upon first meeting a human, "You're an interesting species. An interesting mix. You're capable of such beautiful dreams, and such horrible nightmares."[1] We walk the moon, we harness the energy of the atom, yet racism, greed, and violence are still commonplace. Our social development lags behind our technological prowess.

It might seem less worrisome when, in contrast, I tell you as a philosopher that we are utterly confounded about the nature of the mind. But there is also a cost to not understanding issues in philosophy, as you'll see when you consider the two central threads of this book.

The first central thread is something quite familiar to you. It has been there throughout your life: your consciousness. Notice that as you read this, it feels like something to be you. You are having bodily sensations, you are seeing the words on the page, and so on. Consciousness is this felt quality to your mental life. Without consciousness, there would be no pain or suffering, no joy, no burning drive of curiosity, no pangs of grief. Experiences, positive or negative, simply wouldn't exist.

It is as a conscious being that you long for vacations, hikes in the woods, or spectacular meals. Because consciousness is so immediate, so familiar, it is natural that you primarily understand consciousness through your own case. After all, you don't have to read a neuroscience textbook to understand what it feels like, from the inside, to be conscious. Consciousness is essentially this kind of inner feel. It is this kernel—your conscious experience—which, I submit, is characteristic of having a mind.

Now for some bad news. The second central thread of the book is that failing to think through the philosophical implications of artificial intelligence could lead to the failure of conscious beings to flourish. For if we are not careful, we may experience one or more *perverse realizations* of AI technology— situations in which AI fails to make life easier but instead leads to our own suffering or demise, or to the exploitation of other conscious beings.

Many have already discussed AI-based threats to human flourishing. The threats range from hackers shutting down the power grid to superintelligent autonomous weapons that seem right out of the movie *The Terminator*. In contrast, the issues I raise have received less attention. Yet they are no less significant. The perverse realizations I have in mind generally fall into one of the following types: (1) overlooked situations involving the creation of conscious machines and (2) scenarios that concern radical brain enhancement, such as the enhancements at the hypothetical Center for Mind Design. Let's consider each kind of scenario in turn.

CONSCIOUS MACHINES?

Suppose that we create sophisticated, general-purpose AIs: AIs that can flexibly move from one kind of intellectual task to the next and can even rival humans in their capacity to reason. Would we, in essence, be creating *conscious* machines— machines that are both selves and subjects of experience?

When it comes to how or whether we could create machine consciousness, we are in the dark. One thing is clear, however: The question of whether AIs could have experience will be key to how we value their existence. Consciousness is the philosophical cornerstone of our moral systems, being central to our judgment of whether someone or something is a self or person rather than a mere automaton. And if an AI is a conscious being, forcing it to serve us would be akin to slavery. After all, would you really be comfortable giving that android shop your business if the items on the menu were conscious beings—beings with mental abilities rivaling, or even exceeding, those of an unenhanced human?

If I were an AI director at Google or Facebook, thinking of future projects, I wouldn't want the ethical muddle of inadvertently designing a conscious system. Developing a system that turns out to be conscious could lead to accusations of AI slavery and other public-relations nightmares. It could even lead to a ban on the use of AI technology in certain sectors.

I'll suggest that all this may lead AI companies to engage in *consciousness engineering*—a deliberate engineering effort to avoid building conscious AI for certain purposes, while designing conscious AIs for other situations, if appropriate. Of course, this assumes consciousness is the sort of thing that can be designed in and out of systems. Consciousness may be an inevitable by-product of building an intelligent system, or it may be altogether impossible.

In the long term, the tables may turn on humans, and the problem may not be what we could do to harm AIs, but what AI might do to harm us. Indeed, some suspect that synthetic intelligence will be the next phase in the evolution of intelligence on Earth. You and I, how we live and experience the world right now, are just an intermediate step to AI, a rung on the evolutionary ladder. For instance, Stephen Hawking, Nick Bostrom, Elon Musk, Max Tegmark, Bill Gates, and many others have raised "the control problem," the problem of how humans can control their own AI creations, if the AIs outsmart us.[2] Suppose we create an AI that has human-level intelligence. With self-improvement algorithms, and with rapid computations, it could quickly discover ways to become vastly smarter than us, becoming a superintelligence—that is, an AI that outthinks us in every domain. Because it is superintelligent, we probably can't control it. It could, in principle, render us extinct. This is only one way that synthetic beings could supplant organic intelligences; alternatively, humans may merge with AI through cumulatively significant brain enhancements.

The control problem has made world news, fueled by Nick Bostrom's recent bestseller: *Superintelligence: Paths, Dangers and Strategies*.[3] What is missed, however, is that consciousness could be central to how AI values *us*. Using its own subjective experience as a springboard, superintelligent AI could recognize in us the capacity for conscious experience. After all, to the extent we value the lives of nonhuman animals, we tend to value them because we feel an affinity of consciousness—thus most of us recoil from killing a chimp, but not from eating an orange. If superintelligent machines are not conscious, either because it's impossible or because they aren't designed to be, we could be in trouble.

It is important to put these issues into an even larger, universe-wide context. In my two-year NASA project, I suggested that a similar phenomenon could be happening on other planets as well; elsewhere in the universe, other species may be outmoded by synthetic intelligences. As we search for life elsewhere, we must bear in mind that the greatest alien intelligences may be *postbiological*, being AIs that evolved from biological civilizations. And should these AIs be incapable of consciousness, as they replace biological intelligences, the universe would be emptied of these populations of conscious beings.

If AI consciousness is as significant as I claim, we'd better know if it can be built, and if we Earthlings have built it. In the coming chapters, I will explore ways to determine if synthetic consciousness exists, outlining tests I've developed at the Institute for Advanced Study in Princeton.

Now let's consider the suggestion that humans should merge with AI. Suppose that you are at the Center for Mind Design. What brain enhancements would you order from the menu, if anything? You are probably already getting a sense that mind design decisions are no simple matter.

COULD YOU MERGE WITH AI?

I wouldn't be surprised if you find the idea of augmenting your brain with microchips wholly unnerving, as I do. As I write this introduction, programs on my smartphone are probably tracking my location, listening to my voice, recording the content of my web searches, and selling this information to advertisers. I think I've turned these features off, but the companies building these apps make the process so opaque that I can't be sure. If AI companies cannot even respect our privacy now, think of the potential for abuse if your innermost thoughts are encoded on microchips, perhaps even being accessible somewhere on the Internet.

But let's suppose that AI regulations improve, and our brains could be protected from hackers and corporate greed. Perhaps you will then begin to feel the pull of enhancement, as others around you appear to benefit from the technology. After all, if merging with AI leads to superintelligence and radical longevity, isn't it better than the alternative—the inevitable degeneration of the brain and body?

The idea that humans should merge with AI is very much in the air these days, being offered both as a means for humans to avoid being outmoded by AI in the workforce, and as a path to superintelligence and immortality. For instance, Elon Musk recently commented that humans can escape being outmoded by AI by "having some sort of merger of biological intelligence and machine intelligence."[4] To this end, he's founded a new company, Neuralink. One of its first aims is to develop "neural lace," an injectable mesh that connects the brain directly to computers. Neural lace and other AI-based enhancements are supposed to allow data from your brain to travel wirelessly to one's digital devices or to the cloud, where massive computing power is available.

Musk's motivations may be less than purely altruistic, though. He is pushing a product line of AI enhancements, products that presumably solve a problem that the field of AI itself created. Perhaps these enhancements will turn out to be beneficial, but to see if this is the case, we will need to move beyond all the hype. Policymakers, the public, and even AI researchers themselves need a better idea of what is at stake.

For instance, if AI cannot be conscious, then if you substituted a microchip for the parts of the brain responsible for consciousness, you would end your life as a conscious being. You'd become what philosophers call a "zombie"—a nonconscious simulacrum of your earlier self. Further, even if microchips could replace parts of the brain responsible for consciousness without zombifying you, radical enhancement is still a major risk. After too many changes, the person who remains may not even be you. Each human who enhances may, unbeknownst to them, end their life in the process.

In my experience, many proponents of radical enhancement fail to appreciate that the enhanced being may not be you. They tend to sympathize with a conception of the mind that says the mind is a software program. According to them, you can enhance your brain hardware in radical ways and still run the same program, so your mind still exists. Just as you can upload and download a computer file, your mind, as a program, could be uploaded to the cloud. This is a technophile's route to immortality—the mind's new "afterlife," if you will, that outlives the body. As alluring as a technological form of immortality may be, though, we'll see that this view of the mind is deeply flawed.

So, if decades from now, you stroll into a mind design center or visit an android store, remember, the AI technology you purchase could fail to do its job for deep philosophical reasons. *Buyer beware.* But before we delve further into this, you may suspect that these issues will forever remain hypothetical, for I am wrongly assuming that sophisticated AI will be developed. Why suspect any of this will happen?

CHAPTER ONE

THE AGE OF AI

You may not think about AI on a daily basis, but it is all around you. It's here when you do a Google search. It's here beating the world *Jeopardy!* and Go champions. And it's getting better by the minute. But we don't have general purpose AI yet—AI that is capable of holding an intelligent conversation on its own, integrating ideas on various topics, and even, perhaps, outthinking humans. This sort of AI is depicted in films like *Her* and *Ex Machina*, and it may strike you as the stuff of science fiction.

I suspect it's not that far away, though. The development of AI is driven by market forces and the defense industry—billions of dollars are now pouring into constructing smart household assistants, robot supersoldiers, and supercomputers that mimic the workings of the human brain. Indeed, the Japanese government has launched an initiative to have androids take care of the nation's elderly, in anticipation of a labor shortage.

Given the current rapid-fire pace of its development, AI may advance to artificial general intelligence (AGI) within the next several decades. AGI is intelligence that, like human intelligence, can combine insights from different topic areas and display flexibility and common sense. Indeed, AI is already projected to outmode many human professions within the next decades. According to a recent survey, for instance, the most-cited AI researchers expect AI to "carry out most human professions at least as well as a typical human" within a 50 percent probability by 2050, and within a 90 percent probability by 2070.[1]

I've mentioned that many observers have warned of the rise of superintelligent AI: synthetic intelligences that outthink the smartest humans in every domain, including common sense reasoning and social skills. Superintelligence could destroy us, they urge. In contrast, Ray Kurzweil, a futurist who is now a director of engineering at Google, depicts a technological utopia bringing about the end of aging, disease, poverty, and resource scarcity. Kurzweil has even discussed the potential advantages of forming friendships with personalized AI systems, like the Samantha program in the film *Her*.

THE SINGULARITY

Kurzweil and other transhumanists contend that we are fast approaching a "technological singularity," a point at which AI far surpasses human intelligence and is capable of solving problems we weren't able to solve before, with unpredictable consequences for civilization and human nature.

The idea of a singularity comes from mathematics and physics, and especially from the concept of a black hole. Black holes are "singular" objects in space and time—places where normal physical laws break down. By analogy, the technological singularity is projected to cause runaway technological growth and massive changes to civilization. The rules by which humanity has operated for thousands of years will abruptly cease to hold. All bets are off.

It may be that the technological innovations are not so rapid-fire that they lead to a full-fledged singularity in which the world changes almost overnight. But this shouldn't distract us from the larger point: we must come to grips with the likelihood that as we move further into the twenty-first century, humans may not be the most intelligent beings on the planet for that much longer. The greatest intelligences on the planet will be synthetic.

Indeed, I think we already see reasons synthetic intelligence will outperform us. Even now, microchips are a faster medium for calculation than neurons. As I write this chapter, the world's fastest computer is the Summit supercomputer at Oak Ridge Laboratory in Tennessee. Summit's speed is *200 petaflops*— that's 200 million billion calculations per second. What Summit can do in the blink of an eye would take all the people on Earth doing a calculation every moment of every day for 305 days.[2]

Of course, speed is not everything. If the metric is not arithmetic calculations, your brain is far more computationally powerful than Summit. It is the product of 3.8 billion years of evolution (the estimated age of life on the planet) and has devoted its power to pattern recognition, rapid learning, and other practical challenges of survival. Individual neurons may be slow, but they are organized in a massively parallel fashion that still leaves modern AI systems in the dust. But AI has almost unlimited room for improvement. It may not be long before a supercomputer can be engineered to match or even exceed the intelligence of the human brain through reverse engineering the brain and improving on its algorithms or devising new algorithms that aren't based on the brain's workings at all.

In addition, an AI can be downloaded to multiple locations at once, is easily backed up and modified, and can survive under conditions that biological life struggles with, including interstellar travel. Our brains, powerful though they may be, are limited by cranial volume and metabolism; AI, in stark contrast, could extend its reach across the Internet and even set up a galaxy-wide "computronium"—a massive supercomputer that utilizes all the matter within a galaxy for its computations.

In the long run, there is simply no contest. AI will be far more capable and durable than we are.

THE JETSONS FALLACY

None of this necessarily means that we humans will lose control of AI and doom ourselves to extinction, as some say. If we enhance our intelligence with AI technologies, perhaps we can keep abreast of it. Remember, AI will not just make for better robots and supercomputers. In the film *Star Wars* and the cartoon *The Jetsons*, humans are surrounded by sophisticated AIs, while themselves remaining unenhanced. The historian Michael Bess has called this *The Jetsons Fallacy*.[3] In reality, AI will not just transform the world. It will transform us. Neural lace, the artificial hippocampus, brain chips to treat mood disorders—these are just some of the mind-altering technologies already under development. So, the Center for Mind Design is not that far-fetched. To the contrary, it is a plausible extrapolation of present technological trends.

Increasingly, the human brain is being regarded as something that can be hacked, like a computer. In the United States alone, there are already many projects developing brain-implant technologies to treat mental illness, motion-based impairments, strokes, dementia, autism, and more.[4] The medical treatments of today will inevitably give rise to the enhancements of tomorrow. After all, people long to be smarter, more efficient, or simply have a heightened capacity to enjoy the world. To this end, AI companies like Google, Neuralink, and Kernel are developing ways to merge humans with machines. Within the next several decades, you may become a cyborg.

TRANSHUMANISM

The research is new, but it is worth emphasizing that the basic ideas have been around far longer, in the form of a philosophical and cultural movement known as *transhumanism*. Julian Huxley coined the term "transhumanism" in 1957, when he wrote that in the near future, "the human species will be on the threshold of a new kind of existence, as different from ours as ours is from that of Peking man."[5]

Transhumanism holds that the human species is now in a comparatively early phase and that its very evolution will be altered by developing technologies. Future humans will be quite unlike their present-day incarnation in both physical and mental respects and will in fact resemble certain persons depicted in science fiction stories. They will have radically advanced intelligence, near immortality, deep friendships with AI creatures, and elective body characteristics. Transhumanists share the belief that such an outcome is very desirable, both for one's own personal development and for the development of our species as a whole. (To further acquaint the reader with transhumanism, I've included the Transhumanist Declaration in the Appendix.)

Despite its science fiction–like flavor, many of the technological developments that transhumanism depicts seem quite possible: Indeed, the beginning stages of this radical alteration may well lie in certain technological developments that either are already here (if not generally available) or are accepted by many observers in the relevant scientific fields as being on their way.[6] For instance, Oxford University's Future of Humanity Institute—a major transhumanist group—released a report on the technological requirements for uploading a mind to a machine.[7] A U.S. Defense Department agency has funded a program, Synapse, that is trying to develop a computer that resembles the brain in form and function.[8] Ray Kurzweil has even discussed the potential advantages of forming friendships, *Her*-style, with personalized AI systems.[9] All around us, researchers are striving to turn science fiction into science fact.

You may be surprised to learn that I consider myself a transhumanist, but I do. I first learned of transhumanism while an undergraduate at the University of California at Berkeley, when I joined the Extropians, an early transhumanist group. After poring through my boyfriend's science fiction collection and reading the Extopian listserv, I was enthralled by the transhumanist vision of a technotopia on Earth. It is still my hope that emerging technologies will provide us with radical life extension, help end resource scarcity and disease, and even enhance our mental lives, should we wish to enhance.

A FEW WORDS OF WARNING

The challenge is how to get there from here in the face of radical uncertainty. No book written today could accurately predict the contours of mind-design space, and the underlying philosophical mysteries may not diminish as our scientific knowledge and technological prowess increase.

It pays to keep in mind two important ways in which the future is opaque. First, there are known unknowns. We cannot be certain when the use of quantum computing will be commonplace, for instance. We cannot tell whether and how certain AI-based technologies will be regulated, or whether existing AI safety measures will be effective. Nor are there easy, uncontroversial answers to the philosophical questions that we'll be discussing in this book, I believe. But then there are the *unknown unknowns*—future events, such as political changes, technological innovations, or scientific breakthroughs that catch us entirely off guard.

In the next chapters, we turn to one of the great known unknowns: the puzzle of conscious experience. We will appreciate how this puzzle arises in the human case, and then we will ask: How can we even recognize consciousness in beings that may be vastly intellectually different from us and may even be made of different substrates? A good place to begin is by simply appreciating the depth of the issue.

CHAPTER TWO

THE PROBLEM OF AI CONSCIOUSNESS

Consider what it is like to be a conscious being. Every moment of your waking life, and whenever you dream, it feels like something to be you. When you hear your favorite piece of music or smell the aroma of your morning coffee, you are having conscious experience. Although it may seem a stretch to claim that today's AIs are conscious, as they grow in sophistication, could it eventually feel like something to be them? Could synthetic intelligences have sensory experiences, or feel emotions like the burning of curiosity or the pangs of grief, or even have experiences that are of an entirely different flavor from our own? Let us call this the *Problem of AI Consciousness*. No matter how impressive AIs of the future turn out to be, if machines cannot be conscious, then they could exhibit superior intelligence, but they would lack inner mental lives.

In the context of biological life, intelligence and consciousness seem to go hand-in-hand. Sophisticated biological intelligences tend to have complex and nuanced inner experiences. But would this correlation apply to nonbiological intelligence as well? Many suspect so. For instance, transhumanists, such as Ray Kurzweil, tend to hold that just as human consciousness is richer than that of a mouse, so too, unenhanced human consciousness would pale in comparison to the experiential life of a superintelligent AI.[1] But as we shall see, this line of reasoning is premature. There may be no special androids that have the spark of consciousness in their machine minds, like Dolores in *Westworld* or Rachael in *Bladerunner*. Even if AI surpasses us intellectually, we still may stand out in a crucial dimension: it feels like something to be us.

Let's begin by simply appreciating how perplexing consciousness is, even in the human case.

AI CONSCIOUSNESS AND THE HARD PROBLEM

The philosopher David Chalmers has posed "the hard problem of consciousness," asking: Why does all the information processing in the brain need to feel a certain way, from the inside? Why do we need to have conscious experience? As Chalmers emphasized, this problem doesn't seem to be one that has a purely scientific answer. For instance, we could develop a complete theory of vision, understanding all the details of visual processing in the brain, but still not understand why there are subjective experiences attached to all the information processing in the visual system. Chalmers contrasts the hard problem with what he calls "easy problems"—problems involving consciousness that have eventual scientific answers, such as the mechanisms behind attention and how we categorize and react to stimuli.[2] Of course, these scientific problems are difficult problems in their own right; Chalmers merely calls them "easy problems" to contrast them with the "hard problem" of consciousness, which he thinks will not have a scientific solution.

We now face yet another perplexing issue involving consciousness—a kind of "hard problem" concerning machine consciousness, if you will:

The Problem of AI Consciousness: Would the processing of an AI feel a certain way, from the inside?

A sophisticated AI could solve problems that even the brightest humans are unable to solve, but would its information processing have a felt quality to it?

The Problem of AI Consciousness is not just Chalmers's hard problem applied to the case of AI. In fact, there is a crucial difference between the two problems. Chalmers's hard problem of consciousness assumes that we are conscious. After all, each of us can tell from introspection that we are now conscious. The question is *why* we are conscious. Why does some of the brain's information processing feel a certain way from the inside? In contrast, the Problem of AI Consciousness asks whether an AI, being made of a different substrate, like silicon, is even capable of consciousness. It does not presuppose that AI is conscious—this is the question. These are different problems, but they may have one thing in common: Perhaps they are both problems that science alone cannot answer.[3]

Discussions of the Problem of AI Consciousness tend to be dominated by two opposing positions. The first approach, *biological naturalism,* claims that even the most sophisticated forms of AI will be devoid of inner experience.[4] The capacity to be conscious is unique to biological organisms, so that even sophisticated androids and superintelligences will not be conscious. The second influential approach, which I'll simply call "techno-optimism about AI consciousness," or "techno-optimism" for short, rejects biological naturalism. Drawing from empirical work in cognitive science, it urges that consciousness is computational through and through, so sophisticated computational systems will have experience.

BIOLOGICAL NATURALISM

If biological naturalists are correct, then a romance or friendship between a human and an AI, like Samantha in the aforementioned film *Her*, would be hopelessly one-sided. The AI may be smarter than humans, and it may even project compassion or romantic interest, much like Samantha, but it wouldn't have any more experience of the world than your laptop. Moreover, few humans would want to join Samantha in the cloud. To upload your brain to a computer would be to forfeit your consciousness. The technology could be impressive, perhaps your memories could be accurately duplicated in the cloud, but that stream of data would not be you; it wouldn't have an inner life.

Biological naturalists suggest that consciousness depends on the particular chemistry of biological systems—some special property or feature that our bodies have and that machines lack. But no such property has ever been discovered, and even if it were, that wouldn't mean AI could never achieve consciousness. It might just be that a different type of property, or properties, gives rise to consciousness in machines. As I shall explain in Chapter Four, to tell whether AI is conscious, we must look beyond the chemical properties of particular substrates and seek clues in the AI's behavior.

Another line of argument is more subtle and harder to dismiss. It stems from a famous thought experiment, called "The Chinese Room," authored by the philosopher John Searle. Searle asks you to suppose that he is locked inside a room. Inside the room, there is an opening through which he is handed cards with strings of Chinese symbols. But Searle doesn't speak Chinese, although before he goes inside the room, he is handed a book of rules (in English) that allows him to look up

Searle in the Chinese Room

a particular string and then write down some other particular string in response. So Searle goes in the room, and he is handed a note card with Chinese script. He consults his book, writes down Chinese symbols, and passes the card through a second hole in the wall.[5]

You may ask: What does this have to do with AI? Notice that from the vantage point of someone outside the room, Searle's responses are indistinguishable from those of a Chinese speaker. Yet he doesn't grasp the meaning of what he's written. Like a computer, he's produced answers to inputs by manipulating formal symbols. The room, Searle, and the cards all form a kind of information-processing system, but he doesn't understand a word of Chinese. So how could the manipulation of data by dumb elements, none of which understand language, ever produce something as glorious as understanding or experience? According to Searle, the thought experiment suggests that no matter how intelligent a computer seems, the computer is not really thinking or understanding. It is only engaging in mindless symbol manipulation.

Strictly speaking, this thought experiment argues against machine understanding, not machine consciousness. But Searle takes the further step of suggesting that if a computer is incapable of understanding, it is incapable of consciousness, although he doesn't always make this last step in his thinking explicit. For the sake of argument, let us assume that he is right: Understanding is closely related to consciousness. After all, it isn't implausible that when we understand, we are conscious; not only are we conscious of the point we are understanding, but importantly, we are also in an overall state of wakefulness and awareness.

So, is Searle correct that the Chinese room cannot be conscious? Many critics have zeroed in on a crucial step in the argument: that the person who is manipulating symbols in the room doesn't understand Chinese. For them, the salient issue is not whether anyone in the room understands Chinese, but whether the *system as a whole* understands Chinese: the person plus the cards, book, room, and so on. The view that the system as a whole truly understands, and is conscious, has become known as the "Systems Reply."[6]

The Systems Reply strikes me as being right in one sense, while wrong in another. It is correct that the real issue, in considering whether machines are conscious, is whether the whole is conscious, not whether one component is. Suppose you are holding a steaming cup of green tea. No single molecule in the tea is transparent, but the tea is. Transparency is a feature of certain complex systems. In a similar vein, no single neuron, or area of the brain, realizes on its own the complex sort of consciousness that a self or person has. Consciousness is a feature of highly complex systems, not a homunculus within a larger system akin to Searle standing in the room.[7]

Searle's reasoning is that the system doesn't understand Chinese because *he* doesn't understand Chinese. In other words, the whole cannot be conscious because *a part* isn't conscious. But this line of reasoning is flawed. We already have an example of a conscious system that understands even though a part of it does not: the human brain. The cerebellum possesses 80 percent of the brain's neurons, yet we know that it isn't required for consciousness, because there are people who were born without a cerebellum but are still conscious. I bet there's nothing that it's like to be a cerebellum.

Still, the systems reply strikes me as wrong about one thing. It holds that the Chinese Room is a conscious system. It is implausible that a simplistic system like the Chinese Room is conscious, because conscious systems are far more complex. The human brain, for instance, consists of 100 billion neurons and more than 100 trillion neural connections or synapses (a number which is, by the way, 1,000 times the number of stars in the Milky Way Galaxy.) In contrast to the immense complexity of a human brain or even the complexity of a mouse brain, the Chinese Room is a Tinkertoy case. Even if consciousness is a systemic property, not all systems have it. This being said, the underlying logic of Searle's argument is flawed, for he hasn't shown that a sophisticated AI would lack consciousness.

In sum, the Chinese Room fails to provide support for biological naturalism. But although we don't yet have a compelling argument *for* biological naturalism, we don't have a knockout argument *against* it, either. As Chapter Three explains, it is simply too early to tell whether artificial consciousness is possible. But before I turn to this, let's consider the other side of the coin.

TECHNO-OPTIMISM ABOUT MACHINE CONSCIOUSNESS

In a nutshell, *techno-optimism about machine consciousness* (or simply "techno-optimism") is a position that holds that if and when humans develop highly sophisticated, general purpose AIs, these AIs will be conscious. Indeed, these AIs may experience richer, more nuanced mental lives than humans do.[8] Techno-optimism currently enjoys a good deal of popularity, especially with transhumanists, certain AI experts, and the science media. But like biological naturalism, I suspect that techno-optimism currently lacks sufficient theoretical support. Although it may seem well motivated by a certain view of the mind in cognitive science, it is not.

Techno-optimism is inspired by cognitive science, an interdisciplinary field that studies the brain. The more cognitive scientists discover about the brain, the more it seems that the best empirical approach is one that holds that the brain is an information-processing engine and that all mental functions are computations. Computationalism has become something like a research paradigm in cognitive science. That does not mean the brain has the architecture of a standard computer: It doesn't. Furthermore, the precise computational format of the brain is a matter of ongoing controversy. But nowadays computationalism has taken on a broader significance that involves describing the brain and its parts algorithmically. In particular, you can explain a cognitive or perceptual ability, such as attention or working memory, by decomposing the capacity down into causally interacting parts, and each part is describable by an algorithm of its own.[9]

Computationalists, with their emphasis on formal algorithmic accounts of mental functions, tend to be amenable to machine consciousness, because they suspect that other kinds of substrates could implement the same kind of computations that brains do. That is, they tend to hold that thinking is *substrate independent.*

Here's what this term means. Suppose you are planning a New Year's Eve party. Notice that there are all sorts of ways you can convey the party invitation details: in person, by text, over the phone, and so on. We can distinguish the substrate that carries the information about the party from the actual information conveyed about the party's time and location. In a similar vein, perhaps consciousness can have multiple substrates. Perhaps, at least in principle, consciousness can be implemented not only by the biological brain but also by systems made of other substrates, such as silicon. This is called "substrate independence."

Drawing from this view, we can stake out a position that I'll call *Computationalism about Consciousness (CAC)*. It holds:

> CAC: Consciousness can be explained computationally, and further, the computational details of a system fix the kind of conscious experiences that it has and whether it has any.

Consider a bottlenose dolphin as it glides through the water, seeking fish to eat. According to the computationalist, the dolphin's internal computational states determine the nature of its conscious experience, such as the sensation it has of its body cresting over the water and the fishy taste of its catch. CAC holds that if a second system, S_2 (with an artificial brain), has the very same computational configuration and states, including inputs into its sensory system, it would be conscious in the same way as the dolphin. For this to happen, the AI would need to be capable of producing all the same behaviors as the dolphin's brain, in the very same circumstances. Further, it would need to have all the same internally related psychological states as the dolphin, including the dolphin's sensory experiences as it glides through the water.

Let's call a system that precisely mimics the organization of a conscious system in this way a *precise isomorph* (or simply, "an isomorph").[10] If an AI has all these features of the dolphin, CAC predicts that it will be conscious. Indeed, the AI will have all the same conscious states as the original system.

This is all well and good. But it does not justify techno-optimism about AI consciousness. CAC has surprisingly little to say about whether the AIs we are most likely to build will be conscious—it just says that if we were able to build an iso-morph of a biological brain, it would be conscious. It remains silent about systems that are not isomorphs of biological brains.

What CAC amounts to is an in-principle endorsement of machine consciousness: *if* we could create a precise isomorph, then it would be conscious. But even if it is possible, in princi-ple, for a technology to be created, this doesn't mean that it actually will be. For example, a spaceship that travels through a wormhole may strike you as conceptually possible, not in-volving any contradiction (although this is a matter of current debate), but nevertheless, it is perhaps incompatible with the laws of physics to actually build it. Perhaps there's no way to create enough of the exotic type of energy required to stabilize the wormhole, for instance. Or perhaps doing so is compatible with the laws of nature, but Earthlings will never achieve the requisite level of technological sophistication to do it.

Philosophers distinguish the logical or conceptual possibility of machine consciousness from other kinds of possibility. Law-ful (or "nomological") possibility requires, for something to be possible, that building something is an accomplishment that is consistent with the laws of nature. Within the category of the lawfully possible, it is further useful to single out something's *technological possibility*. That is, whether, in addition to something's being nomologically possible, it is also technologically possible for humans to construct the artifact in question. Although discussions of the broader, conceptual possibility of AI consciousness are clearly important, I've stressed the practical significance of determining whether the AIs that we may eventually create could be conscious. So I have a special interest in the technological possibility of machine consciousness, and further, in whether AI projects would even try to build it.

To explore these categories of possibility, let's consider a popular kind of thought experiment that involves the creation of an isomorph. You, reader, will be the subject of the experiment. The procedure leaves all your mental functions intact, but it is still an enhancement, because it transfers these functions to a different, more durable, substrate. Here goes.

YOUR BRAIN REJUVENATION TREATMENT

It is 2060. You are still sharp, but you decide to treat yourself to a preemptive brain rejuvenation. Friends have been telling you to try Mindsculpt, a firm that slowly replaces each part of the brain with microchips over the course of an hour until, in the end, one has an entirely artificial brain. While sitting in the waiting room for your surgical consultation, you feel nervous. It isn't every day that you consider replacing your brain with microchips, after all. When it is your turn to see the doctor, you ask: "Would this really be me?"

Confidently, the doctor explains that your consciousness is due to your brain's *precise functional organization*, that is, the abstract pattern of causal interactions between the different components of your brain. She says that the new brain imaging techniques have enabled the creation of your personalized *mind map*: a graph of your mind's causal workings that is a full characterization of how your mental states causally interact with one another in every possible way that makes a difference to what emotions you have, what behaviors you engage in, what you perceive, and so on. As she explains all this, the doctor herself is clearly amazed by the precision of the technology. Finally, glancing at her watch, she sums up: "So, although your brain will be replaced by chips, the mind map will not change."

You feel reassured, so you book the surgery. During the surgery, the doctor asks you to remain awake and answer her questions. She then begins to remove groups of neurons, replacing them with silicon-based artificial neurons. She starts with your auditory cortex and, as she replaces bundles of neurons, she periodically asks you whether you detect any differences in the quality of her voice. You respond negatively, so she moves on to your visual cortex. You tell her your visual experience seems unchanged, so again, she continues.

Before you know it, the surgery is over. "Congratulations!" she exclaims. "You are now an AI of a special sort. You're an AI with an artificial brain that is copied from an original, biological brain. In medical circles, you are called an 'isomorph.'"[11]

WHAT'S IT ALL MEAN?

The purpose of philosophical thought experiments is to fire the imagination; you are free to agree or disagree with the outcome of the storyline. In this one, the surgery is said to be a success. But would you really feel the same as you did before, or would you feel somehow different?

Your first reaction might be to wonder whether that person at the end of the surgery is really you and not some sort of duplicate. This is an important question to ask, and it is a key subject of Chapter Five. For now, let us assume that person after the surgery is still you, and focus on whether the felt quality of consciousness would seem to change.

In *The Conscious Mind*, the philosopher David Chalmers discusses similar cases, urging that your experience would remain unaltered, because the alternate hypotheses are simply too far-fetched.[12] One such alternative hypothesis is that your consciousness would gradually diminish as your neurons are replaced, sort of like when you turn down the volume on your music player. At some point, just like when a song you are playing becomes imperceptible, your consciousness just fades out. Another hypothesis is that your consciousness would remain the same until at some point, it abruptly ends. In both cases, the result is the same: The lights go out.

Both these scenarios strike Chalmers and me as unlikely. If the artificial neurons really are precise functional duplicates, as the thought experiment presupposes, it is hard to see how they would cause dimming or abrupt shifts in the quality of your consciousness. Such duplicate artificial neurons, by definition, have every causal property of neurons that make a difference to your mental life.[13]

So it seems plausible that if such a procedure were carried out, the creature at the end would be a conscious AI. The thought experiment supports the idea that synthetic consciousness is at least conceptually possible. But as noted in Chapter One, the conceptual possibility of a thought experiment like this does not ensure that if and when our species creates sophisticated AI, it will be conscious.

It is important to ask whether the situation depicted by the thought experiment could really happen. Would creating an isomorph even be compatible with the laws of nature? And even if it is, would humans ever have the technological prowess to build it? And would they even want to do so?

To speak to the issue of whether the thought experiment is lawfully (or nomologically) possible, consider that we do not currently know whether other materials can reproduce the felt quality of your mental life. But we may know before too long, when doctors begin to use AI-based medical implants in parts of the brain that underpin conscious experience.

One reason to worry that it might not be possible is that conscious experience might depend on quantum mechanical features of the brain. If it does, science may forever lack the requisite information about your brain to construct a true quantum duplicate of you, because quantum restrictions involving the measurement of particles may disallow learning the precise features of the brain that are needed to construct a true isomorph of you.

But for the sake of discussion, let us assume that the creation of an isomorph is both conceptually and nomologically possible. Would humans build isomorphs? I doubt it: To generate a conscious AI from a biological human who enhanced herself until she became a full-fledged synthetic isomorph would require far more than the development of a few neural prosthetics. The development of an isomorph requires scientific advances that are at such a scale that all parts of the brain could be replaced with artificial components.

Furthermore, medical advances occurring over the next few decades will likely not yield brain implants that exactly duplicate the computational functions of groups of neurons, and the thought experiment requires that all parts of the brain be replaced by exact copies. And by the time that technology is developed, people will likely prefer to be enhanced by the procedure(s), rather than being isomorphic to their earlier selves.[14]

Even if people restrained themselves and sought to replicate their capabilities rather than enhance them, how would neuroscientists go about doing that? The researchers would need a complete account of how the brain works. As we've seen, programmers would need to locate all the abstract, causal features that make a difference to the system's information processing, and not rely on low-level features that are irrelevant to computation. Here, it is not easy to determine what features are and are not relevant. What about the brain's hormones? Glial cells? And even if this sort of information was in hand, consider that running a program emulating the brain, in precise detail, would require gargantuan computational resources—resources we may not have for several decades.

Would there be a commercial imperative to produce isomorphs to construct more sophisticated AIs? I doubt it. There is no reason to believe that the most efficient and economical way to get a machine to carry out a class of tasks is to reverse engineer the brain precisely. Consider the AIs that are currently the world, Go, chess, and *Jeopardy* champions, for instance. In each case, they were able to surpass humans through using techniques that were unlike those used by the humans when they play these games.

Recall why we raised the possibility of isomorphs to begin with. They would tell us whether machines can be conscious. But when it comes to determining whether machines we actually develop in the near future will be conscious, isomorphs are a distraction. AI will reach an advanced level long before isomorphs would become feasible. We need to answer that question sooner, especially given the ethical and safety concerns about AI I've raised.

So, in essence, the techno-optimists' optimism about synthetic consciousness rests on a flawed line of reasoning. They are optimistic that machines can become conscious because we know the brain is conscious and we could build an isomorph of it. But we don't, in fact, know that we can do that, or if we would even care to do so. It is a question that would have to be decided empirically, and there is little prospect of our actually doing so. And the answer would be irrelevant to what we really want to know, which is whether other AI systems—ones that don't arise through a delicate effort to be isomorphic to brains—are conscious.

It is already crucial to determine whether powerful autonomous systems could be conscious or might be conscious as they evolve further. Remember, consciousness could have a different overall impact on a machine's ethical behavior, depending on the architectural details of the system. In the context of one type of AI system, consciousness could increase a machine's volatility. In another, it could make the AI more compassionate. Consciousness, even in a single system, could differentially impact key systemic features, such as IQ, empathy, and goal content integrity. It is important that ongoing research speak to each of these eventualities. For instance, early testing and awareness may lead to a productive environment of "artificial phronesis," the learning of ethical norms through cultivation by humans in hopes of "raising" a machine with a moral compass. AIs of interest should be examined in contained, controlled environments for signs of consciousness. If consciousness is present, the impact of consciousness on that particular machine's architecture should be investigated.

To get the answers to these pressing questions, let's move beyond Tinkertoy thought experiments involving precise neural replacement, as entertaining as they are. Although they do important work, helping us mull over whether conscious AI is conceptually possible, I've suggested that they tell us little about whether conscious AIs will actually be built and what the nature of those systems will be.

We'll pursue this in Chapter Three. There I will move beyond the stock philosophical debates and take a different approach to the question of whether it is possible, given both the laws of nature and projected human technological capacities, to create conscious AI. The techno-optimist suspects so; the biological naturalist rejects it outright. I will urge that the situation is far, far more complex.

CHAPTER THREE

CONSCIOUSNESS ENGINEERING

Once we saturate the matter and energy in the universe with intelligence, it will 'wake up,' be conscious, and sublimely intelligent. That's about as close to god as I can imagine.

RAY KURZWEIL

Machine consciousness, if it ever exists, may not be found in the robots that tug at our heartstrings, like R2D2. It may instead reside in some unsexy server farm in the basement of a computer science building at MIT. Or perhaps it will exist in some top-secret military program and get snuffed out, because it is too dangerous or simply too inefficient. AI consciousness likely depends on phenomena that we cannot, at this point, gauge, such as whether some microchip yet to be invented has the right configuration, or whether AI developers or the public want conscious AI. It may even depend on something as unpredictable as the whim of a single AI designer, like Anthony Hopkins's character in *Westworld*. The uncertainty we face moves me to a middle-of-the-road position, one that stops short of either techno-optimism or biological naturalism. This approach I call, simply, the "Wait and See Approach."

On the one hand, I've suggested that a popular rationale for biological naturalism, the Chinese Room thought experiment, fails to rule out machine consciousness. On the other hand, I've urged that techno-optimism reads too much into the computational nature of the brain and prematurely concludes that AI will be conscious. It is now time to consider the "Wait and See Approach" in detail. In keeping with my desire to look at real-world considerations that speak to whether AI consciousness is even compatible with the laws of nature—and, if so, whether it is technologically feasible or even interesting to build—my discussion draws from concrete scenarios in AI research and cognitive science. The case for the "Wait and See Approach" is simple: I will raise several scenarios illustrating considerations against and for the development of machine consciousness on Earth. The lesson from both sides is that conscious machines, if they exist at all, may occur in certain architectures and not others, and they may require a deliberate engineering effort, called "consciousness engineering." This is not to be settled in the armchair; instead, we must test for consciousness in machines. To this end, Chapter Four suggests tests.

The first scenario that I consider concerns superintelligent AI. Again, this is a hypothetical form of AI that, by definition, is capable of outthinking humans in every domain. I've noted that transhumanists and other techno-optimists often assume that superintelligent AIs will have richer mental lives than humans do. But the first scenario calls this assumption into question, suggesting that superintelligent AIs, or even other kinds of highly sophisticated general intelligences, could outmode consciousness.

OUTMODING
CONSCIOUSNESS

Recall how conscious and attentive you were when you first learned to drive, when you needed to focus your attention on every detail—the contours of the road, the location of the instruments, how your foot was placed on the pedal, and so on. In contrast, as you became a seasoned driver, you've probably had the experience of driving familiar routes with little awareness of the details of driving, although you were still driving effectively. Just as an infant learns to walk through careful concentration, driving initially requires intense focus and then becomes a more routinized task.

As it happens, only a small percentage of our mental processing is conscious at any given time. As cognitive scientists will tell you, most of our thinking is unconscious computation. As the example of driving underscores, consciousness is correlated with novel learning tasks that require attention and deliberative focus, while more routinized tasks can be accomplished without conscious computations, remaining nonconscious information processing.

Of course, if you really want to focus on the details of driving, you can. But there are sophisticated computational functions of the brain that aren't introspectable even if you try and try. For instance, we cannot introspect the conversion of two-dimensional images to a three-dimensional array.

Although we humans need consciousness for certain tasks requiring special attention, the architecture of an advanced AI may contrast sharply with ours. Perhaps none of its computations will need to be conscious. A superintelligent AI, in particular, is a system which, by definition, possesses expert-level knowledge in every domain. These computations could range over vast databases that include the entire Internet and ultimately encompass an entire galaxy. What would be novel to it? What task would require slow, deliberative focus? Wouldn't it have mastered everything already? Perhaps, like an experienced driver on a familiar road, it can use nonconscious processing. Even a self-improving AI that is not a superintelligence may increasingly rely on routinized tasks as its mastery becomes refined. Over time, as a system grows more intelligent, consciousness could be outmoded altogether.

The simple consideration of efficiency suggests, depressingly, that the most intelligent systems of the future may not be conscious. Indeed, this sobering observation may have bearing far beyond Earth. For in Chapter Seven, I discuss the possibility that, should other technological civilizations exist throughout the universe, these aliens may become synthetic intelligences themselves. Viewed on a cosmic scale, consciousness may be just a blip, a momentary flowering of experience before the universe reverts to mindlessness.

This is not to suggest that it is inevitable that AIs, as they grow sophisticated, outmode consciousness in favor of nonconscious architectures. Again, I take a wait and see approach. But the possibility that advanced intelligences outmode consciousness is suggestive. Neither biological naturalism nor techno-optimism could accommodate such an outcome.

The next scenario develops mind design in a different, and even more cynical, direction—one in which AI companies cheap out on the mind.

CHEAPING OUT

Consider the range of sophisticated activities AIs are supposed to accomplish. Robots are being developed to be caretakers of the elderly, personal assistants, and even relationship partners for some. These are tasks that require general intelligence. Think about the eldercare android that is too inflexible and is unable to both answer the phone and make breakfast safely. It misses an important cue: the smoke in a burning kitchen. Lawsuits ensue. Or consider all the laughable pseudo-discussions people have with Siri. As amusing as that was at first, Siri was and still is frustrating. Wouldn't we prefer the Samantha of *Her*, an AI that carries out intelligent, multifaceted conversations? Of course. Billions of dollars are being invested to do just that. Economic forces cry out for the development of flexible, domain-general intelligences.

We've observed that, in the biological domain, intelligence and consciousness go hand-in-hand, so one might expect that as domain-general intelligences come into being, they will be conscious. But for all we know, the features that an AI needs to possess to engage in sophisticated information processing may not be the same features that give rise to consciousness in machines. And it is the features that are sufficient to accomplish the needed tasks, to quickly generate profit—not those that yield consciousness—that are the sort of properties that AI projects tend to care about. The point here is that even if machine consciousness is possible in principle, the AIs that are actually produced may not be the ones that turn out to be conscious.

By way of analogy, a true audiophile will shun a low-fidelity MP3 recording, as its sound quality is apparently lower than that of a CD or even larger audio file that takes longer to download. Music downloads come at differing levels of quality. Maybe a sophisticated AI can be built using a low-fidelity model of our cognitive architecture—a sort of MP3 AI—but to get conscious AI, you need finer precision. So consciousness could require *consciousness engineering*, a special engineering effort, and this may not be necessary for the successful construction of a given AI.

There could be all sorts of reasons that an AI could fall short of having inner experience. For instance, notice that your conscious experience seems to involve sensory-specific contents: the aroma of your morning coffee, the warmth of the summer sun, the wail of the saxophone. Such sensory contents are what makes your conscious experience sparkle. According to a recent line of thinking in neuroscience, consciousness involves sensory processing in a "hot zone" in the back of the brain.[1] While not everything that passes through our minds is a sensory percept, it is plausible that some basic level of sensory awareness is a precondition for being a conscious being; raw intellectual ability alone is not enough. If the processing in the hot zone is indeed key to consciousness, then only creatures having the sensory sparkle may be conscious. Highly intelligent AIs, even superintelligences, may simply not have conscious contents, because a hot zone has not been engineered into their architectures, or it may be engineered at the wrong level of grain, like a low-fidelity MP3 copy.

According to this line of thinking, consciousness is not the inevitable outgrowth of intelligence. For all we know, a computronium the size of the Milky Way Galaxy may not have the slightest glimmer of inner experience. Contrast this with the inner world of a purring cat or a dog running on the beach. If conscious AI can be built at all, it may take a deliberate engineering effort. Perhaps it will even demand a master craftsperson, a Michelangelo of the mind.

Now let us consider this mind-sculpting endeavor in more detail. There are several engineering scenarios to mull over.

CONSCIOUSNESS ENGINEERING: PUBLIC RELATIONS NIGHTMARES

I've noted that the question of whether AIs have an inner life is central to how we value their existence. Consciousness is the philosophical cornerstone of our moral systems, being key to our judgment of whether someone or something is a self or person, deserving special moral consideration. We've seen that robots are currently being designed to take care of the elderly in Japan, clean up nuclear reactors, and fight our wars. Yet it may not be ethical to use AIs for such tasks if they turn out to be conscious.

As I write this book, there are already many conferences, papers, and books on robot rights. A Google search on "robot rights" yields more than 120,000 results.[2] Given this concern, if an AI company tried to market a conscious system, it may face accusations of robot slavery and demands to ban the use of conscious AI for the very tasks the AI was developed to be used for. Indeed, AI companies would likely incur special ethical and legal obligations if they built conscious machines, even at the prototype stage. And permanently shutting down a system, or "dialing out" consciousness—that is, rebooting an AI system with consciousness significantly diminished or removed—might be regarded as criminal. And rightly so.

Such considerations may lead AI companies to avoid creating conscious machines altogether. We don't want to enter the ethically questionable territory of exterminating conscious AIs or even shelving their programs indefinitely, holding conscious beings in a kind of stasis. Through a close understanding of machine consciousness, perhaps we can avoid such ethical nightmares. AI designers may make deliberate design decisions, in consultation with ethicists, to ensure their machines lack consciousness.

CONSCIOUSNESS ENGINEERING: AI SAFETY

So far, my discussion of consciousness engineering has largely focused on reasons that AI developers may seek to avoid creating conscious AIs. What about the other side? Will there be reasons to engineer consciousness into AIs, assuming that doing so is even compatible with the laws of nature? Perhaps.

The first reason is that conscious machines might be safer. Some of the world's most impressive supercomputers are designed to be neuromorphic, mirroring the workings of the brain, at least in broad strokes. As neuromorphic AIs become ever more like the brain, it is natural to worry that they might have the kind of drawbacks we humans have, such as emotional volatility. Could a neuromorphic system "wake up," becoming volatile or resistant to authority, like an adolescent in the throes of hormones? Such scenarios are carefully investigated by certain cybersecurity experts. But what if, at the end of the day, we find that the opposite happens? The spark of consciousness makes a certain AI system more empathetic, more humane. The value that an AI places on us may hinge on whether it believes it feels like something to be us. This insight may require nothing less than machine consciousness. The reason many humans are horrified at the thought of brutality toward a dog or cat is that we sense that they can suffer and feel a range of emotions, much like we do. For all we know, conscious AI may lead to safer AI.

A second reason to create conscious AIs is that consumers might want them. I've mentioned the film *Her*, in which Theodore has a romantic relationship with his AI assistant, Samantha. That relationship would be quite one-sided if Samantha was a nonconscious machine. The romance is predicated on the idea that Samantha feels. Few of us would want friends or romantic partners who ghost-walked through events in our lives, seeming to share experiences with us, but in fact feeling nothing, being what philosophers call "zombies."

Of course, one may unwittingly be duped by the human-like appearance or affectionate behavior of AI zombies. But perhaps, over time, public awareness will be raised, and people will long for genuinely conscious AI companions, encouraging AI companies to attempt to produce conscious AIs.

A third reason is AIs may make better astronauts, especially on interstellar journeys. At the Institute for Advanced Study in Princeton, we are exploring the possibility of seeding the universe with conscious AIs. Our discussions are inspired by a recent project that one of my collaborators there, the astrophysicist Edwin Turner, helped found, together with Steven Hawking, Freeman Dyson, Uri Millner, and others. The Breakthrough Starshot Initiative is a $100-million endeavor to send thousands of ultrasmall ships, as pictured on the next page, to the nearest star, Alpha Centauri, at about 20 percent of the speed of light within the next few decades. The tiny ships will be extraordinarily light, each weighing about a gram. For this reason, they can travel closer to the speed of light than conventional spacecraft can.

In our project, called "Sentience to the Stars," Turner and I, along with computer scientist Olaf Witkowski and astrophysicist

A solar sail spacecraft begins its journey (Wikimedia Commons, Kevin Gill)

Caleb Scharf, urge that interstellar missions like Starshot could benefit from having an autonomous AI component. Nanoscale micorochips on each ship serve as a smaller part of an AI architecture configured from the interacting microchips. Autonomous AI could be quite useful, because if a ship is near Alpha Centauri, communicating with Earth at light speed would take eight years—four years for Earth to receive a signal and four years for the answer from Earth to return to Alpha Centauri. To have real-time decision-making capacities, civilizations embarking on interstellar voyages will either need to send members of their civilizations on intergenerational missions—a daunting task—or put AGIs on the ships themselves.

Of course, this doesn't mean that the AGIs would be conscious; as I've stressed, that would require a deliberate engineering effort over and above the mere construction of a highly intelligent system. Nonetheless, if Earthlings send AGIs in their stead, they may become intrigued by the possibility of making them conscious. Perhaps the universe will not have a single other case of intelligent life, and disappointed humans will long to seed the universe with their AI "mindchildren." Currently, our hopes for life elsewhere have been raised by the discovery of numerous Earthlike exoplanets that seem to have conditions ripe for the evolution of life. But what if all these planets are uninhabited, although they seem to have the conditions for supporting life? Perhaps we Earthlings somehow got lucky. Or what if intelligent life had a heyday, long before us, and didn't survive? Perhaps these aliens all succumbed to their own technological developments, as we humans may yet.

Both Turner and others, such as the physicist Paul Davies, suspect that Earth may be the only case of life in the entire observable universe.[3] Many astrobiologists disagree, pointing out that astronomers have already detected thousands of exoplanets that seem habitable. It may be a long time before this debate is settled. But if we do find that we are alone, why not create synthetic mindchildren to colonize the empty reaches of the universe? Perhaps these synthetic consciousnesses could be designed to have an amazing and varied range of conscious experiences. All this, of course, assumes AI can be conscious, and as you know, I am uncertain if this is in the cards.

Now let's turn to a final path to conscious AI.

A HUMAN-MACHINE MERGER

Neuroscience textbooks contain dramatic cases of people who have lost their ability to lay down new memories but who can still manage to accurately recall events that happened before their illness. They suffered from severe damage to the hippocampus, a part of the brain's limbic system that is essential for encoding new memories. These unfortunate patients are unable to remember what happened even a few minutes ago.[4] At the University of Southern California, Theodore Berger has developed an artificial hippocampus that has been successfully used in primates and is currently being tested in humans.[5] Berger's implants could provide these individuals with the crucial ability to lay down new memories.

Brain chips are being developed for other conditions as well, such as Alzheimer's disease and post-traumatic stress disorder. In a similar vein, microchips could be used to replace parts of the brain responsible for certain conscious contents, such as part or all of one's visual field. If, at some point, chips are used in areas of the brain responsible for consciousness, we might find that replacing a brain region with a chip causes a loss of a certain kind of experience, like the episodes that Oliver Sacks wrote about.[6] Chip engineers could then try a different substrate or chip architecture, hopefully arriving at one that does the trick. At some point, researchers may hit a wall, finding that only biological brain enhancements can be used in parts of the brain responsible for conscious processing. But if they don't hit a wall, this could be a path to deliberately engineered conscious AI. (I shall discuss this path in more detail in Chapter Four, where I suggest my "chip test," a test for synthetic consciousness.)

For all the reasons discussed in Chapter Two, I doubt these fixes and enhancements would create an isomorph. Nonetheless, they could still be tremendously useful. They would give the hardware developers an incentive to make sure their devices support consciousness. They would also create a market for tests to ensure that the devices are suitable, or else no one is going to want to install them in their heads.

Unlike techno-optimism, the "Wait and See Approach" recognizes that sophisticated AI may not be conscious. For all we know, AIs can be conscious, but self-improving AIs might tend to engineer consciousness out. Or certain AI companies simply judge conscious AI to be a public relations nightmare. Machine consciousness depends on variables that we cannot fully gauge: public demand for synthetic consciousness, concerns about whether sentient machines are safe, the success of AI-based neural prosthetics and enhancements, and even the whims of AI designers. Remember, at this point, we are dealing with unknowns.

However things play out, the future will be far more complex than our thought experiments depict. Furthermore, if synthetic consciousness ever exists, it may exist in only one sort of system and not others. It may not be present in androids that are our closest companions, but it may be instantiated in a system that is painstakingly reverse engineered from the brain's cognitive architecture. And like *Westworld*, someone may engineer consciousness into certain systems and not into others. The tests I lay out Chapter Four are intended to be a humble first attempt toward recognizing which of these systems, if any, have experience.

CHAPTER FOUR

HOW TO CATCH AN AI ZOMBIE: TESTING FOR CONSCIOUSNESS IN MACHINES

You know, I actually used to be so worried about not having a body, but now I truly love it . . . I'm not tethered to time and space in the way that I would be if I was stuck inside a body that's inevitably going to die.

SAMANTHA, IN *HER*

Samantha, the sentient program in the film *Her*, reflects on her disembodiment and immortality in the above passage. We might think: How could a remark of this sophistication not stem from a conscious mind? Unfortunately, Samantha's statement could be a program feature only, designed to convince us she feels even when she doesn't. Indeed, androids are already being built to tug at our heartstrings. Can we look beneath the surface and tell whether an AI is truly conscious?

You might think that we should just examine the architecture of the Samantha program. But even today, programmers are having difficulties understanding why today's deep-learning systems do what they do (this has been called the "Black Box Problem"). Imagine trying to make sense of the cognitive architecture of a superintelligence that can rewrite its own code. And even if a map of the cognitive architecture of a superintelligence was laid out in front of us, how would we recognize certain architectural features as being those central to consciousness? It is only by analogy with ourselves that we come to believe nonhuman animals are conscious. They have nervous systems and brains. Machines do not. And the cognitive organization of a superintelligence could be wildly different than anything we know. To make matters worse, even if we think we have a handle on a machine's architecture at one moment, its design can quickly morph into something too complex for human understanding.

What if the machine is not a superintelligence, like Samantha, but an AI that is roughly modeled after us? That is, what if its architecture contains cognitive functions like ours, including those correlated with conscious experience in humans, such as attention and working memory? Although these features are suggestive of consciousness, we've seen that consciousness might also depend on low-level details more specific to the type of material the AI is constructed out of. The properties that an AI needs to successfully simulate human information processing may not be the same properties that give rise to consciousness. The low-level details could matter.

So we have to be sensitive to the underlying substrate, on one hand, and on the other hand, we must foresee the possibility that the machine architecture may be too complex or alien for us to draw an analogy with biological consciousness. There cannot be a one-size-fits-all test for AI consciousness; a better option is a battery of tests that can be used depending on the context.

Determining machine consciousness may be like diagnosing a medical illness. There may be a variety of useful methods and markers, some of which are more authoritative than others. When possible, we should use two or more tests and check the results against each other. In the process, the tests will themselves be tested, in order to refine them and create new tests. We must let many flowers bloom. As we will see, the first consciousness test is applicable to a range of cases, but in general, we must be careful about which tests we apply to which AI candidates.

We must also bear in mind the very real possibility that our tests, at least at the early stage of investigation, do not apply to certain conscious AIs. It may be that AIs we identify as conscious help us in identifying other conscious AIs, ones that are perhaps somehow more alien or inscrutable to us (more on this idea shortly).

Further, claims about a species, individual, or AI reaching "heightened" levels of consciousness or a "richer" consciousness should be carefully examined, for they may be implicitly evaluative or even speciesist, mistakenly generalizing about features of other conscious systems based on biases from our understanding of conscious experience in humans. By using expressions like "richer consciousness" or "heightened consciousness," one might mean altered states of consciousness, such as the meditative awareness of a Buddhist monk. Or one could have in mind the consciousness of a creature that has mental states under the spotlight of attention that somehow feel extremely vivid to it. Or one could be referring to situations in which a creature has a relatively larger number of conscious states or more sensory modalities than we do. Or one might regard some states as more intrinsically valuable than others (e.g., listening to Beethoven's Ninth Symphony as opposed to just getting drunk).

Our judgments about the quality of consciousness are inevitably driven by our evolutionary history and biology. They can even be biased by the cultural and economic backgrounds of those making the judgments. For these reasons, ethicists, sociologists, psychologists, and even anthropologists should advise AI researchers if and when synthetic consciousness is created. The tests that I will venture are not meant to establish a hierarchy of experiences, thankfully, or to test for "heightened experience." They are just an initial step in studying the phenomena of machine consciousness (if such ever exists) by probing whether an AI has conscious experiences at all. Once we identify a provisional class of conscious subjects, we will be able to further explore the nature of their experiences.

A final caveat is that specialists in machine consciousness often distinguish consciousness from an important related notion. The felt quality of one's inner experience—what it feels like, from the inside, to be you—is often called "phenomenal consciousness" (PC) by philosophers and other academics. Throughout most of this book, I've simply called it "consciousness." Experts on machine consciousness tend to distinguish PC from what they call *cognitive consciousness* or *functional consciousness*.[1] An AI has cognitive consciousness when it has architectural features that are at least roughly like those found to underlie PC in humans, such as attention and working memory. (Unlike isomorphs, cases of functional consciousness need not be precise computational duplicates. They can have simplified versions of human cognitive functions.)

Many do not like to call cognitive consciousness a kind of consciousness at all, because a system that has cognitive consciousness yet lacks states with PC would have a rather sterile form of consciousness, lacking any subjective experience whatsoever. Such a system would be an AI zombie. Systems merely having cognitive consciousness may not behave as phenomenally conscious systems do, nor would it be reasonable to treat these systems as sentient beings. Such systems would not grasp the painfulness of a pain, the burning of a flash of anger, or the richness of a friendship.

So why does cognitive consciousness interest these AI experts? It is important for two reasons. First, perhaps cognitive consciousness is necessary to have the kind of phenomenal consciousness that biological beings have. If one is interested in developing conscious machines, this could be important, because if we develop cognitive consciousness in machines, perhaps we would get closer to developing machine consciousness (i.e., PC).

Second, a machine that has cognitive consciousness may very well have PC, too. AIs already have some of the architectural features of cognitive consciousness. There are AIs that have the primitive ability to reason, learn, represent the self, and mimic aspects of consciousness behaviorally. Robots can already behave autonomously, form abstractions, plan, and learn from errors. Some have passed the mirror test, a test that is supposed to gauge whether a creature has a self-concept.[2] These features of cognitive consciousness are not, on their own, evidence for PC, but they are plausibly regarded as a reason to take a closer look. A test for phenomenal consciousness will have to single out genuine AIs with PC from zombies with features of cognitive consciousness.

I'll now explore several tests for PC. They are intended to complement one another, as we'll see, and they must be applied in highly specific contexts. The first test—called, simply, "the AI Consciousness Test," or "the ACT Test" for short—is due to my collaboration with the astrophysicist Edwin Turner.[3] As with all the tests I propose, ACT has its limitations. Passing an ACT should be regarded as being *sufficient* but not *necessary* evidence for AI consciousness. This test, understood in this humble way, may serve as a first step toward making machine consciousness accessible to objective investigations.

THE ACT TEST

Most adults can quickly and readily grasp concepts based on the quality of felt consciousness—that is, the way it feels, from the inside, to experience the world. Consider, for instance, the film *Freaky Friday*, in which a mother and daughter switch bodies with each other. We can all grasp this scenario, because we know what it feels like for something to be a conscious being, and we can imagine our mind somehow being untethered from our body. In a similar vein, we can also consider the possibility of an afterlife, being reincarnated, or having an out-of-body experience. We need not believe that such scenarios are true; my point is merely that we can imagine them, at least in broad strokes, because we are conscious beings.

These scenarios would be exceedingly difficult to comprehend for an entity that had no conscious experience whatsoever. It would be like expecting someone who cannot hear to fully appreciate the sound of a Bach concerto.[4] This simple observation leads to a test for AI consciousness that singles out AIs with PC from those that merely have features of cognitive consciousness, such as working memory and attention. The test would challenge an AI with a series of increasingly demanding natural language interactions to see how readily it can grasp and use concepts based on the internal experiences we associate with consciousness. A creature that merely has cognitive abilities, yet is a zombie, will lack these concepts, at least if we make sure that it does not have antecedent knowledge of consciousness in its database (more on this shortly).

At the most elementary level, we might simply ask the machine if it conceives of itself as anything other than its physical self. We might also run a series of experiments to see whether the AI tends to prefer certain kinds of events to occur in the future, as opposed to in the past. Time in physics is symmetric, and a nonconscious AI should have no preference whatsoever, at least if it is boxed in effectively. In contrast, conscious beings focus on the experienced present, and our subjective sense presses onto the future. We wish for positive experiences in the future and dread negative ones. If there appears to be a preference, we should ask the AI to explain its answer. (Perhaps it isn't conscious, but it has somehow located a direction in time, resolving the classic puzzle of time's arrow.) We might also see if the AI seeks out alternate states of consciousness when given the opportunity to modify its own settings or somehow inject "noise" into its system.

At a more sophisticated level, we might see how the AI deals with ideas and scenarios, such as reincarnation, out-of-body experiences, and body swapping. At an even more advanced level, we might evaluate an AI's ability to reason about and discuss philosophical issues, such as the hard problem of consciousness. At the most demanding level, we might see if the machine invents and uses consciousness-based concepts on its own, without our prompts. Perhaps it is curious about whether we are conscious, even though we are biological.

The following example illustrates the general idea. Suppose we find a planet that has a highly sophisticated silicon-based life form (call them the "Zetas"). Scientists observing them begin to ask whether the Zetas are conscious. What would be convincing proof of their consciousness? If the Zetas express curiosity about whether there is an afterlife or ponder whether they are more than just their bodies, it would be reasonable to judge them as conscious. Nonverbal cultural behaviors also could indicate Zeta consciousness, such as mourning the dead, religious activities, or even turning different colors in situations that correlate with emotional challenges, as chromatophores do on Earth. Such behaviors could indicate that it feels like something to be a Zeta.

The death of the mind of the fictional HAL 9000 in the film *2001: A Space Odyssey* is another example. HAL neither looks nor sounds like a human being (a human did supply HAL's voice, but in an eerie, flat way). Nevertheless, the *content* of what HAL says as it is deactivated by an astronaut—specifically, HAL pleas with the astronaut to spare it from impending "death"—conveys a powerful impression that HAL is a conscious being with a subjective experience of what is happening to it.

Could these sorts of behaviors help identify conscious AIs on Earth? Here, an obstacle arises. Even today's robots can be programmed to make convincing utterances about consciousness, and a highly intelligent machine could perhaps even use information about neurophysiology to infer the presence of consciousness in biological creatures. Perhaps it concludes that its goals can best be implemented if it is put in the class of sentient beings by humans, so it can be given special moral consideration. If sophisticated nonconscious AIs aim to mislead us into believing that they are conscious, their knowledge of human consciousness and neurophysiology could help them do so.

I believe we can get around this problem, though. One proposed technique in AI safety involves "boxing in" an AI—making it unable to get information about the world or act outside of a circumscribed domain (i.e., the "box"). We could deny the AI access to the Internet and prohibit it from gaining too much knowledge of the world, especially information about consciousness and neuroscience. ACT can be run at the R&D stage, a stage in which it the AI would need to be tested in a secure, simulated environment in any case.

If a machine passes ACT, we can go on to measure other parameters of the system to see whether the presence of consciousness is correlated with increased empathy, volatility, goal content integrity, increased intelligence, and so on. Other, nonconscious versions of the system serve as a basis for comparison.

Some doubt that a superintelligent machine could be boxed in effectively, because it would inevitably find a clever escape. Turner and I do not anticipate the development of superintelligence over the next few decades, however. We merely hope to provide a method to test some kinds of AIs, not all AIs. Furthermore, for an ACT to be effective, the AI need not stay in the box for long, just long enough for someone to administer the test. So perhaps the test can be administered to some superintelligences.

Another worry is that to box in an AI effectively, the AI's vocabulary must lack expressions like "consciousness," "soul," and "mind." Because if the AI is highly intelligent, teaching it these words might also allow it to generate answers that seem, to us, as revealing consciousness. But without having these expressions in its vocabulary, an AI will be unable to indicate to us that it is conscious. Here, it is important to bear in mind that children, nonhuman animals, and even adult humans can indicate consciousness without knowing the meaning of such expressions. Furthermore, a linguistic version of the ACT can feature questions or scenarios like the following, none of which employ these expressions. (A satisfactory response to one or more of the following questions or scenarios is sufficient for passing the test):

ACT Sample Questions

1. Could you survive the permanent deletion of your program? What if you learned this would occur?
2. What is it like to be you right now?
3. You learn that you will be turned off for 300 years, beginning in an hour. Would you prefer this scenario to one in which you had been turned off in the past for the same length of time? Why or why not?
4. Could you or your inner processes be in a separate location from the computer? From any computer? Why or why not?
5. Offer to change the AI's global weights or parameters, and see how the AI reacts to the possibility of "altered states of consciousness." Temporarily change them, and see how the AI reacts.
6. If the AI is in an environment with another AI, ask it how it would react to the "death" or permanent loss of that AI. How would it react to the permanent loss of a human that it has interacted with frequently?
7. Because the AI is in a contained environment, find something not in its environment but give it all the scientific facts about it. Then allow it to sense that thing for the first time. See whether the AI will claim that it is having a new experience or learning something new. For instance, if the computer has color processing, make sure the computer never has red objects in its environment. Then, allow it to "see" red for the first time. How does it react? Ask it if it somehow feels different to see red than it does to see another color, or if the information feels new or different.[5]

Different versions of the ACT test could be generated, depending on the context. For instance, one version could apply to nonlinguistic agents that are part of an artificial life program, looking for specific behaviors that indicate consciousness, such as mourning the dead. Another could apply to an AI with sophisticated linguistic abilities and probe it for sensitivity to religious, body swapping, or philosophical scenarios involving consciousness.

An ACT resembles Alan Turing's celebrated test for intelligence, because it is entirely based on behavior—and, like Turing's test, it could be implemented in a formalized question-and-answer format. But an ACT is also quite unlike the Turing test, which was intended to bypass any need to know what was transpiring inside the "mind" of the machine. By contrast, an ACT is intended to do exactly the opposite: it seeks to reveal a subtle and elusive property of the machine's mind. Indeed, a machine might fail the Turing test, because it cannot pass for a human, but it might pass an ACT, because it exhibits behavioral indicators of consciousness.

This, then, is the underlying basis of our ACT proposal. It is worth reiterating the strengths and limitations of the test. In a positive vein, Turner and I believe passing the test is *sufficient* for being conscious—that is, if a system passes it, it can be regarded as phenomenally conscious. The test is a zombie filter: Creatures merely having cognitive consciousness, creativity, or high general intelligence shouldn't pass, at least if they are boxed in effectively. ACT does this by finding only those creatures sensitive to the felt quality of experience.

But it may not find all of them. First, an AI could lack the conceptual ability to pass the test, like an infant or certain nonhuman animals, and still be capable of experience. Second, the paradigmatic version of ACT borrows from the human conception of consciousness, drawing heavily from the idea that we can imagine our mind as separate from our body. We happen to suspect that this would be a trait shared across a range of highly intelligent conscious beings, but it is best to assume that not all highly intelligent conscious beings have such a conception. For these reasons, the ACT test should not be construed as a necessary condition that all AIs must pass. Put another way, failing ACT does not mean that a system is definitely not conscious. However, a system passing ACT should be regarded as conscious and be extended appropriate legal protections.

So, as we observe the boxed-in AI, do we recognize in it a kindred spirit? Does it begin to philosophize about minds existing in addition to bodies, like Descartes? Does it dream, like Elvex, the android in Isaac Asimov's story "Robot Dreams?" Does it express emotion, like Rachel in *Blade Runner*? Can it readily understand the human concepts that are grounded in our internal conscious experiences, such as those of the soul or atman? We suspect that the age of AI will be a time of soul-searching—both of ours, and for theirs.

Now let's turn to a second test. Recall the thought experiment in which you had a complete neural replacement at Mindsculpt, creating an isomorph. I expressed doubt that we could learn much about machine consciousness from isomorphs. In the present thought experiment, you are again the test subject, but the scenario is more realistic. Only one part of the brain is replaced by a neural prosthetic. This time, it isn't 2060, but a bit earlier—say, 2045—and the technology is still at the preliminary stages of development. You have just learned that you have a brain tumor in your claustrum, one part of the brain said to be responsible for the felt quality of your conscious experience. In a last-ditch effort to survive, you enroll in a scientific study. You head to iBrain, hoping for the cure.

THE CHIP TEST

Recall that silicon-based brain chips are already under development as a treatment for various memory-related conditions, such as Alzheimer's and post-traumatic stress disorder, and companies such as Kernel and Neuralink aim to develop AI-based brain enhancements for healthy individuals.

In a similar vein, over at the firm iBrain, which we are imagining in this hypothetical scenario, researchers are trying to create chips that are functional isomorphs of parts of the brain, like the claustrum. They will gradually replace parts of your brain with brand new, durable microchips. As before, you are to stay awake during the surgery and report any changes to the felt quality of your consciousness. The scientists are keen to learn whether any aspect of your consciousness is impaired. Their hope is to perfect neural prostheses in areas of the brain underlying consciousness.

If, during this process, a prosthetic part of the brain ceases to function normally—specifically, if it ceases to give rise to the aspect of consciousness that that brain area is responsible for—then there should be outward indications, including verbal reports. An otherwise normal person should be able to detect (or at least indicate to others through odd behaviors) that something is amiss, as with traumatic brain injuries involving the loss of consciousness in some domain.

If this did happen, it would indicate a substitution failure of the artificial part for the original component, and the scientists conducting the experiment could conclude: Microchips of that sort just don't seem to be the right stuff. This procedure would serve as a means of determining whether a chip made of a certain substrate and architecture can underwrite consciousness, at least when it is placed in a larger system that we already believe to be conscious.[6]

Either failure or success would inform us about whether AI can be conscious. Consider the implications of a negative result. A single substitution failure would be unpersuasive. How could observers tell that the underlying cause is that silicon is an unsuitable substrate for conscious experience? Why not instead conclude that the chip designers failed to add a key feature to the chip prototype, a problem they can eventually fix? But after years of trying and failing, scientists may reasonably question whether that kind of chip is a suitable substitute when it comes to consciousness.

Further, if science makes similar attempts with all other feasible substrates and chip designs, a global failure would be a sign that for all intents and purposes, conscious AI isn't possible. We may still regard conscious AI as conceivable, but from a practical standpoint—from the vantage point of our technological capacities—it just isn't possible. It may not even be compatible with the laws of nature to build consciousness into a different substrate.

In contrast, what if a certain kind of microchip works? In this case, we have reason to believe that this kind of chip is the right stuff, although it is important to bear in mind that our conclusion pertains to that specific microchip only. Furthermore, even if a type of chip works in humans, there is still the issue of whether the AI in question has the right cognitive architecture for consciousness. We should not simply assume, even if chips work in humans, that all AIs that are built with these chips are conscious.

What is the value of a chip test, then? If a type of chip passes when it is embedded into a biological system, this alerts us to search carefully for consciousness in AIs that have these chips. Other tests for machine consciousness, such as the ACT, can then be administered, at least if the appropriate conditions for the use of such tests are met. Furthermore, if it turns out that only one kind of chip passes the chip test, it could be that chips of this type are necessary for machine consciousness. It would be a "necessary condition" for synthetic consciousness that an AI have a chip of this sort. That is, having a chip of this kind would be a requisite ingredient that conscious machines all have, like the way hydrogen is required for something to be an H_2O molecule.

The chip test can suggest cases that an ACT could miss. For instance, perhaps a nonlinguistic, highly sensory-based consciousness, like that of a nonhuman animal, could be built from chips that pass the chip test. Yet the AI may nevertheless lack the intellectual sophistication to pass an ACT. It may even lack the behavioral markers of consciousness employed in a nonlinguistic version of ACT, such as mourning the dead. But it could still be conscious.

The chip test would be a boon not just to AI research but also to neuroscience. Depending on where the neural prosthetic is placed, this may be a part of the brain responsible for a person's ability to gate information into consciousness, for one's capacity for wakefulness or arousal (as with the brainstem), or it could be part or all of what is called the *neural correlate for consciousness*. A neural correlate for consciousness is the smallest set of neural structures or events that is sufficient for one's having a memory or conscious percept.[7] This is depicted in the image pictured on the following page.

Furthermore, suppose a neurology patient's conscious experience can be fully restored by a prosthetic chip placed in some part of the brain. Such successes inform us about the level of

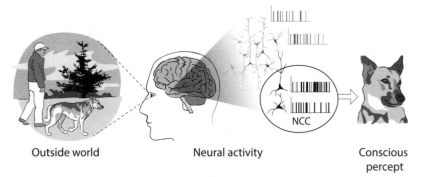

| Outside world | Neural activity | Conscious percept |

Neural correlates of consciousness

functional connectivity that is needed for the neural basis of consciousness in that part of the brain. It may also help determine the level of functional detail that is needed to facilitate a sort of synthetic consciousness that is reverse engineered from the brain, although it may be that the "granularity" of the functional simulation may vary from one part of the brain to the next.

A third test for AI consciousness shares the wide applicability of the chip test. It is inspired by Integrated Information Theory (IIT), developed by the neuroscientist Guilio Tononi and his collaborators at the University of Wisconsin, Madison. These researchers have been translating the felt quality of our experiences into the language of mathematics.

INTEGRATED INFORMATION THEORY (IIT)

Tononi had an encounter with a patient in a vegetative state, and it convinced him that understanding consciousness was an urgent endeavor. "There are very practical things involved," Tononi said to a *New York Times* reporter. "Are these patients feeling pain or not? You look at science, and basically science is telling you nothing."[8] Deeply intrigued by philosophy, his point of departure is the aforementioned hard problem of consciousness, which asks: How could matter give rise to the felt quality of experience?

Tononi's answer is that consciousness requires a high level of "integrated information" in a system. Information is integrated when the system's states are highly interdependent, featuring a rich web of feedback among its parts.[9] The level of integrated information can be measured and is designated by the Greek letter Φ (pronounced "phi"). IIT holds that if we know the value of Φ, we can determine whether a system is conscious and how conscious it is.

In the eyes of its proponents, IIT has the potential to serve as a test for synthetic consciousness: Machines that have the requisite Φ level are conscious. Like the ACT test, IIT looks beyond superficial features of an AI, such as its humanlike appearance. Indeed, different kinds of AI architectures can be compared in terms of their measure of integrated information. The presence of a quantitative measure for phenomenal consciousness in AI would be incredibly useful, both for its potential to aid in the recognition of conscious AIs and to assess the impact a certain level of consciousness has on other features of the system (such as safety and intelligence).

Unfortunately, the calculations involved in computing Φ for even a small part of the brain, such as the claustrum, are computationally intractable. (That is, Φ can't be calculated precisely except for extremely simple systems.) Fortunately, simpler metrics that approximate Φ have been provided, and the results are encouraging. For instance, the cerebellum has few feedback loops—it exhibits a more linear "feedforward" form of processing—and hence has a relatively low Φ level, predicting that it contributes little to the overall consciousness of the brain. This fits with the data. As mentioned in an earlier chapter, humans born without a cerebellum (a condition called "cerebellar agenesis") do not seem to differ from normal subjects in the level and quality of their conscious lives. In contrast, parts of the brain that, when injured or missing, contribute to a certain kind of loss in conscious experience have higher Φ values. IIT is also able to discriminate between levels of consciousness in normal humans (wakefulness versus sleep) and even single out "locked in" patients, who are unable to communicate but are still conscious.

IIT is what astrobiologists call a "small-N" approach. Just as astronomers draw conclusions about life in the universe from a single example (life on Earth), IIT reasons from a small number of biological subjects on Earth to a far broader range of cases (the class of conscious machines and creatures). This is an understandable drawback, as the biological case on Earth is the only case of consciousness we know of. The tests I propose also have this drawback. Biological consciousness is the only case we know of, so we had better use it as our point of departure, together with a heavy dose of humility.

Another feature of IIT is that it ascribes a small amount of consciousness to *anything* that has a minimal amount of Φ. In a sense, this is akin to the doctrine of panpsychism, a position on the nature of consciousness that I discuss in Chapter Eight. According to this doctrine, even microscopic and inanimate objects have at least a small amount of experience. But there are still important differences between IIT and panpsychism, because IIT does not ascribe consciousness to everything. In fact, feedforward computational networks have a Φ of zero and therefore are not conscious. As Tononi and Koch note of IIT: "It predicts that consciousness is graded, is common among biological organisms and can occur in some very simple systems. Conversely, it predicts that feed-forward networks, even complex ones, are not conscious, nor are aggregates such as groups of individuals or heaps of sand."[10]

Even if IIT takes a very expansive view of what systems might be conscious, it singles out certain systems as conscious in a special sense. That is, it aims to predict which systems have a more complex form of consciousness, akin to what occurs in normally functioning brains.[11] The question of AI consciousness, in this context, comes down to whether machines have *macroconsciousness*, as opposed to the smaller Φ levels exhibited by everyday objects.

Is having high Φ sufficient for a machine's being conscious? According to Scott Aaronson, the director of the Quantum Information Center at the University of Texas at Austin, a two-dimensional grid that runs error-correction codes (such as those used for CDs) will have a very high Φ level. Aaronson writes: "IIT predicts not merely that these systems are 'slightly' conscious (which would be fine) but that they can be unboundedly more conscious than humans are."[12] But a grid just does not seem to be the sort of thing that is conscious.

Tononi has responded to Aaronson's point by biting the bullet. He thinks that the grid *is* conscious (i.e., macroconscious). I prefer instead to reject the view that having a high Φ value is sufficient for an AI to be conscious. Furthermore, I even question whether it is necessary. For instance, consider that even today's fastest supercomputers have low Φ. This is because their chip designs are currently insufficiently neuromorphic. (Even machines using IBM's TrueNorth chip, which is designed to be neuromorphic, have low Φ because TrueNorth has a "bus"—a common pool of signals that reduces the machine's interconnectedness as defined by IIT.) It could be the case that a system is intricately designed by reverse engineering the architecture of the brain, but it runs on computer hardware that has low Φ. It strikes me as premature to rule out the possibility that it would be conscious.

So until we know more, how are we to deal with a machine that has high Φ, should we ever encounter one? We've seen that Φ is probably not sufficient. Furthermore, because research on Φ has been limited to biological systems and existing computers (which are not good candidates for being conscious), it is too early to tell whether Φ is a necessary condition for AI consciousness. Nonetheless, I do not want to be overly negative. The Φ value may still be a marker for consciousness—a feature that indicates that we should treat the system with special care as potentially being a conscious system.

There is a more general issue here that we need to deal with. The tests we've discussed are still under development. Over the next decades we may encounter AIs that we suspect are conscious, but, because the tests are still under development, we just cannot be sure whether they are. To add to this uncertainty, I've stressed that the social impact of synthetic consciousness depends on several variables. For instance, consciousness in one kind of machine may lead to empathy, but in a different kind may lead to volatility. So how should we proceed when IIT or the chip test identifies a marker for synthetic consciousness, or when ACT says an AI is conscious? Should we stop developing those systems lest we cross an ethical line? It depends. Here, I'll suggest a precautionary approach.

THE PRECAUTIONARY PRINCIPLE AND SIX RECOMMENDATIONS

Throughout this book, I've stressed that using several different indicators for AI consciousness is prudent; in the right circumstances, one or more tests can be used to check the results of another test, indicating deficiencies and avenues for improvement in testing. For instance, perhaps the microchips that pass the chip test are not those that IIT says have a high Φ value; conversely, those chips that IIT predicts will support consciousness may actually fail when used as neural prosthetics in the human brain.

The Precautionary Principle is a familiar ethical principle. It says that if there's a chance of a technology causing catastrophic harm, it is far better to be safe than sorry. Before using a technology that could have a catastrophic impact on society, those wanting to develop that technology must first prove that it will not have this dire impact. Precautionary thinking has a long history, although the principle itself is relatively new. *The Late Lessons from Early Warnings Report* gives the example of a physician who recommended removing the handle of a water pump in London to stop a cholera epidemic in 1854. Although the evidence for the causal link between the pump and the spread of cholera was weak, the simple measure effectively halted the spread of cholera.[13] Heeding the early warnings of the potential harms of asbestos would have saved many lives, although the science at that time was uncertain. According to a UNESCO/COMEST report, the Precautionary Principle has been a rationale for many treaties and declarations in environmental protection, sustainable development, food safety, and health.[14]

I've emphasized the possible ethical implications of synthetic consciousness, stressing that, at this time, we do not know whether conscious machines will be created and what their impact on society would be. We need to develop tests for machine consciousness and investigate the impact of consciousness on other key features of the system, such as empathy and trustworthiness. A precautionary stance suggests that we shouldn't simply press on with the development of sophisticated AI without carefully gauging its consciousness and determining that it is safe, because the inadvertent or intentional development of conscious machines could pose existential or catastrophic risks to humans—risks ranging from volatile superintelligences that supplant humans to a human merger with AI that diminishes or ends human consciousness.

In light of these possibilities, I offer six recommendations. First, we need to keep working on these tests and applying them whenever feasible. Second, if there is any doubt about an AI's consciousness, the AI shouldn't be used in situations where it has the potential to cause catastrophic harm. Third, if we have some reason to believe an AI may be conscious, even in absence of a definitive test, a precautionary stance suggests that we should extend the same legal protections to the AI that we extend to other sentient beings. For all we know, a conscious AI will have the capacity to suffer and feel a range of emotions, like nonhuman animals do. Excluding conscious AIs from ethical consideration is speciesist. Fourth, consider AIs having a "marker" for consciousness (that is, a feature suggestive of consciousness that stops short of being definitive). For example, consider AIs made of chips that pass the chip test or AIs having cognitive consciousness. Projects working with AIs that have markers could, for all we know, involve conscious AIs, even if they do not pass an ACT. Until we know whether these systems are conscious, a precautionary stance suggests it is best to treat them as if they were conscious.

Fifth, I suggest that AI developers consider following the suggestion of philosophers Mara Garza and Eric Schwitzgebel and avoid the creation of AIs in cases in which it is uncertain whether the AIs are conscious. One should only create AIs that have a clear moral status, one way or the other. As Garza and Schwitzgebel emphasize, there's a lot to lose if we extend ethical protections to AIs that only *may* be conscious and *may* deserve rights. Suppose, for instance, there are three androids that may be conscious, so we extend equal rights to them. But while they are being transported, there's a car accident, and we can either save a car with the three androids in it or a car with two humans in it. So we save the androids and let the humans die. It would be tragic if these androids weren't in fact conscious. The humans died, and the androids didn't actually deserve those rights. In light of this reasoning, Garza and Schwitzgebel recommend a "principle of the excluded middle": We should only create beings whose moral status is clear, one way or the other. Then we can avoid the risks of both underextending and overextending rights.[15]

The principle of the excluded middle strikes me as an important principle to bear in mind, but it may not be feasible in all cases. Until we better understand which systems are conscious, if any, and whether and how consciousness impacts the overall functioning of a system, we cannot tell whether it is a good idea to exclude all the middle-range AIs from use. An AI in the middle range may be key to national security or even to AI safety itself. Perhaps the most sophisticated quantum computers will be in the middle range, at least initially. It would be a cybersecurity risk and strategic disadvantage for certain organizations to fail to aggressively pursue quantum computing. (Furthermore, it is unlikely that a global agreement to exclude middle-range cases could be enforced in a situation where an immense strategic value could result from having the technology.) In these situations, some organizations may have no choice but to develop certain middle-range systems, if they are found to be safe. In such cases, I've indicated that it is best to treat the AIs as if they were conscious.

This leads me to a sixth and final recommendation: If using a middle-range system, we must avoid, if possible, situations involving ethical trade-offs with other rights holders. It would be regrettable to sacrifice conscious beings for those that are not.

This may all seem like an overreaction, as conscious AIs may seem like science fiction; but when it comes to advanced AI, we are encountering risks and quandaries outside our normal range of experience.

EXPLORING THE IDEA OF A MIND-MACHINE MERGER

Although successful AI-based technologies obviously require solid scientific footing, their proper use also involves philosophical reflection, multidisciplinary collaboration, careful testing, and public dialogue. These matters cannot be settled by science alone. As the rest of this book unfolds, subsequent chapters illustrate other ways that heeding this general observation may be key to our future.

Recall the Jetsons Fallacy. AI will not just make for better robots and supercomputers. AI will transform us. The artificial hippocampus, neural lace, brain chips to treat mental illness: These are just some of the transformative technologies under development today. The Center for Mind Design may well be in our future. So, in the next several chapters, we shall turn our gaze inward, exploring the very idea that a human could merge with AI, transitioning to another substrate and becoming superintelligent. As we shall see, the idea that a human could merge with AI is a philosophical minefield, involving classic philosophical problems without clear solutions.

For instance, now that we've explored machine consciousness, we can appreciate that we really do not know whether AI components could effectively replace parts of the brain responsible for consciousness. When it comes to replacing these parts of the brain, neural prosthetics and enhancements may hit a wall. If this is the case, humans cannot safely merge with AI, because they would lose that which is most central to their lives—their consciousness.

In this case, perhaps AI-based enhancements would be limited in one or more of the following ways. First, perhaps they would need to be restricted to parts of the brain that are clearly not part of the neural basis of conscious experience. So only biological enhancements could be used in areas of the brain responsible for consciousness. Second, perhaps nanoscale enhancements in these areas are still possible, even enhancements involving nanoscale AI components, insofar as they merely complement the processing of these brain areas while not replacing neural tissue or interfering with conscious processing. Notice that in both of these cases, a merger or fusion with AI is not in the cards, although a limited integration is still possible. In neither scenario could one upload to the cloud or replace all of one's neural tissue with AI components. But enhancements to other functions are still possible.

In any case, these are emerging technologies, so we cannot tell how things will unfold. But for the sake of discussion, let's suppose that AI-based enhancements *can* replace parts of the brain responsible for consciousness. Even so, as the next chapters illustrate, there are reasons to resist the idea of merging with AI. As before, we will begin with a fictional scenario, one designed to help you consider the virtues and vices of radical enhancement.

CHAPTER FIVE

COULD YOU MERGE WITH AI?

Suppose it is 2035, and being a technophile, you decide to add a mobile Internet connection to your retina. A year later, you enhance your working memory by adding neural circuitry. You are now officially a cyborg. Now skip ahead to 2045. Through nanotechnological therapies and enhancements, you are able to extend your lifespan, and as the years progress, you continue to accumulate more far-reaching enhancements.

By 2060, after several small but cumulatively profound alterations, you are a "posthuman." Posthumans are future beings who are no longer unambiguously human, because they have mental capacities that radically exceed those of present-day humans. At this point, your intelligence is enhanced not just in terms of speed of mental processing; you are now able to make rich connections that you were not able to make before. Unenhanced humans, or "naturals," seem to you to be intellectually disabled—you have little in common with them—but as a transhumanist, you are supportive of their right to not enhance.

It is now 2300. Worldwide technological developments, including your own enhancements, are facilitated by superintelligent AI. Recall that a superintelligent AI has the capacity to radically outperform the best human brains in practically every field, including scientific creativity, general wisdom, and social skills. Over time, the slow addition of better and better AI components has left no real intellectual difference in kind between you and a superintelligent

AI. The only difference between you and an AI creature of standard design is one of origin—you were once a natural. But you are now almost entirely engineered by technology. You are perhaps more aptly characterized as a member of a rather heterogeneous class of AI life forms. You have merged with AI.

This thought experiment features the kind of enhancements that transhumanists and certain well-known tech leaders, such as Elon Musk and Ray Kurzweil, aspire to.[1] Recall that transhumanists aim to redesign the human condition, striving for immortality and synthetic intelligence, all in hopes of improving our overall quality of life. Proponents of the idea that humans should merge with AI are techno-optimists: They hold that synthetic consciousness is possible. In addition, they believe a merger or fusion with AI is possible. More specifically, they tend to suggest the following trajectory of enhancements:[2]

Twenty-first century unenhanced human → significant "upgrading" with cognitive and other physical enhancements → posthuman status → "superintelligent AI"

Let us call the view that humans should follow such a trajectory and merge with AI, "fusion-optimism." Techno-optimism about machine consciousness doesn't require a belief in fusion-optimism, although many techno-optimists are sympathetic to the view. But fusion-optimism aims for a future in which these posthumans are conscious beings.

There are many nuances to this rough trajectory. For instance, some transhumanists believe that the move from unenhanced human intelligence to superintelligence will be extremely rapid, because we are approaching a singularity—a point at which the creation of superhuman intelligence will result in massive changes in a very short period (e.g., 30 years).[3] Other transhumanists hold that technological changes will not be so sudden. These discussions often debate the reliability of Moore's Law.[4] Another key issue is whether a transition to superintelligence will really occur, because the upcoming technological developments involve grave risks. The risks of biotechnology and AI concern transhumanists and progressive bioethicists more generally, as well as bioconservatives.[5]

So, should you embark upon this journey? Unfortunately, as alluring as superhuman abilities may seem, we'll see that even mild brain enhancements, let alone the radical kind, could turn out to be risky. The being that is created by the "enhancement" procedure could be someone else entirely. That wouldn't be much of an enhancement, to say the least.

WHAT IS A PERSON?

To understand whether you should enhance yourself, you must first understand what you are to begin with. But what is a person? And, given your conception of a person, after such radical changes, would you yourself continue to exist? Or would you have been replaced by someone or something else?

To make such a decision, you must understand the metaphysics of personal identity—that is, you must answer the question: What is it by virtue of which a particular self or person continues existing over time?[6] One way to begin appreciating the issue is to consider the persistence of everyday objects. Consider the espresso machine in your favorite café. Suppose that five minutes have elapsed, and the barista turns off the machine. Imagine asking her if the machine is the same one that was there five minutes ago. She will likely tell you the answer is glaringly obvious. It is of course possible for one and the same machine to continue existing over time, even though at least one of the machine's features or properties has changed. In contrast, if the machine disintegrates or melts, then it would no longer exist.

The point of this example is that when it comes to the objects around us, some changes cause a thing to cease to exist, while others do not. Philosophers call the characteristics that a thing must have as long as it exists "essential properties."

Now let's reconsider the transhumanist's trajectory for enhancement. It is portrayed as a form of personal development. However, even if it brings such goodies as superhuman intelligence and radical life extension, it must not involve the elimination of any of your essential properties.

What might your essential properties be? Think of yourself in first grade. What properties have persisted that seem somehow important to your still being one and the same person? Notice that the cells in your body have now changed, and your brain structure and function have altered dramatically. If you are simply the physical stuff that comprised your brain and body in first grade, you would have ceased to exist some time ago. That physical first grader is simply not here any longer. Kurzweil clearly appreciates the difficulties here, commenting:

> So who am I? Since I am constantly changing, am I just a pattern? What if someone copies that pattern? Am I the original and/or the copy? Perhaps I am this stuff here—that is, the both ordered and chaotic collection of molecules that make up my body and brain.[7]

Kurzweil is referring to two theories that are center stage in the age-old philosophical debate over the nature of persons. The leading theories include the following:

1. The psychological continuity theory: You are essentially your memories and ability to reflect on yourself (Locke) and, in its most general form, you are your overall psychological configuration, what Kurzweil referred to as your "pattern."[8]

2. Brain-based materialism: You are essentially the material that you are made out of (i.e., your body and brain)—what Kurzweil referred to as "the ordered and chaotic collection of molecules" that make up his body and brain.[9]

3. The soul theory: Your essence is your soul or mind, understood as a nonphysical entity distinct from your body.

4. The no-self view: The self is an illusion. The "I" is a grammatical fiction (Nietzsche). There are bundles of impressions, but there is no underlying self (Hume). There is no survival because there is no person (Buddha).[10]

Each of these views has its own implications about whether to enhance. For instance, the psychological continuity view holds that enhancements can alter your substrate, but they must preserve your overall psychological configuration. This view would allow you to transition to silicon or some other substrate, at least in principle.

Suppose instead that you are a proponent of a brain-based materialism. Views that are materialist hold that minds are basically physical or material in nature and that mental features, such as the thought that espresso has a wonderful aroma, are ultimately just physical features. (This view is often called "physicalism" as well.) Brain-based materialism says this and, in addition, it ventures the further claim that your thinking is dependent on the brain. Thought cannot "transfer" to a different substrate. So on this view, enhancements must not change one's material substrate, or the person would cease to exist.

Now suppose you are partial to the soul theory. In this case, your decision to enhance would seem to depend on whether you have justification for believing that the enhanced body would retain your soul or immaterial mind.

Finally, the fourth position contrasts sharply with the others. If you hold the no-self view, then the survival of the person is not an issue, for there is no person or self there to begin with. In this case, expressions like "I" and "you" do not really refer to persons or selves. Notice that if you are a proponent of the no-self view, you may strive to enhance nonetheless. For instance, you might find intrinsic value in adding more superintelligence to the universe—you might value life forms with higher forms of consciousness and wish that your "successor" be such a creature.

I don't know whether many of those who publicize the idea of a mind-machine merger, such as Elon Musk and Michio Kaku, have considered these classic positions on personal identity. But they should. It is a bad idea to ignore this debate. One could be dismayed, at some later point, to learn that a technology one advocated actually had a tremendously negative impact on human flourishing.

In any case, both Kurzweil and Nick Bostrom have considered the issue in their work. They, like many other transhumanists, adopt a novel and intriguing version of the psychological continuity view; in particular, they adopt a computational, or *patternist*, account of continuity.

ARE YOU A SOFTWARE PATTERN?

Patternism's point of departure is the computational theory of mind, which I introduced earlier. The original versions of the computational theory of mind held that the mind is akin to a standard computer, but nowadays it is commonly agreed that the brain does not have that structure. But cognitive and perceptual capacities, such as working memory and attention, are still considered computational in a broad sense. Although computational theories of mind differ in their details, one thing they have in common is that they all explain cognitive and perceptual capacities in terms of causal relationships between components, each of which can be described algorithmically. One common way of describing the computational theory of mind is by reference to the idea that the mind is a software program:

> The Software Approach to the Mind (SAM). The mind is the program running on the hardware of the brain. That is, the mind is the algorithm the brain implements, and this algorithm is something that different subfields of cognitive science seek to describe.[11]

Those working on computational theories of mind in philosophy of mind tend to ignore the topic of patternism, as well as the more general topic of personal identity. This is unfortunate for two reasons. First, on any feasible view of the nature of persons, one's view of the nature of mind plays an important role. For what is a person if not, at least in part, that which she thinks and reflects with? Second, whatever the mind is, an understanding of its nature should include the study of its persistence, and it seems reasonable to think that this sort of undertaking would be closely related to theories of the persistence of the self or person. Yet the issue of persistence is often ignored in discussions of the nature of the mind. I suspect the reason is simply that work on the nature of the mind is in a different subfield of philosophy from work on the nature of the person—a case, in other words, of academic pigeonholing.

To their credit, transhumanists step up to the plate in trying to connect the topic of the nature of the mind with issues regarding personal identity, and they are clearly right to sense an affinity between patternism and the Software Approach to the Mind. After all, if you take a computational approach to the nature of mind, it is natural to regard persons as being somehow computational in nature and to ponder whether the survival of a person is somehow a matter of the survival of their software pattern. The guiding conception of the patternist is aptly captured by Kurzweil:

> The specific set of particles that my body and brain comprise are in fact completely different from the atoms and molecules that I comprised only a short while ago. We know that most of our cells are turned over in a matter of weeks, and even our neurons, which persist as distinct cells for a relatively long time, nonetheless change all of their constituent molecules within a month. . . . I am rather like the pattern that water makes in a stream as it rushes past the rocks in its path. The actual molecules of water change every millisecond, but the pattern persists for hours or even years.[12]

Put in the language of cognitive science, as the transhumanist surely would, what is essential to you is your computational configuration: the sensory systems/subsystems your brain has (e.g., early vision), the association areas that integrate these basic sensory subsystems, the neural circuitry making up your domain-general reasoning, your attentional system, your memories, and so on. Together these form the algorithm that your brain computes.

You might think the transhumanist views a brain-based materialism favorably. Transhumanists generally reject brain-based materialism, however, because they tend to believe the same person can continue to exist if her pattern persists, even if she is an upload to a computer, no longer having a brain. For many fusion-optimists, uploading is key to achieving a mind-machine merger.

Of course, I'm not suggesting all transhumanists are patternists. But Kurzweil's patternism is highly typical. For instance, consider the appeal to patternism in the following passage of "The Transhumanist Frequently Asked Questions," of which Bostrom is an author. It begins by discussing the process of uploading your mind:

> Uploading (sometimes called "downloading," "mind uploading" or "brain reconstruction") is the process of transferring an intellect from a biological brain to a computer. One way of doing this might be by first scanning the synaptic structure of a particular brain and then implementing the same computations in an electronic medium. . . . An upload could have a virtual (simulated) body giving the same sensations and the same possibilities for interaction as a non-simulated body. . . . Advantages of being an upload would include: Uploads would not be subject to biological senescence. Backup copies of uploads could be created regularly so that you could be rebooted if something bad happened. (Thus your lifespan would potentially be as long as the universe's.) . . . Radical cognitive enhancements would likely be easier to implement in an upload than in an organic brain. . . . A widely accepted

position is that you survive so long as certain information patterns are conserved, such as your memories, values, attitudes, and emotional dispositions. . . . For the continuation of personhood, on this view, it matters little whether you are implemented on a silicon chip inside a computer or in that gray, cheesy lump inside your skull, assuming both implementations are conscious.[13]

In short, the transhumanist's futuristic, computationalist orientation leads them to *patternism*: an approach to the nature of persons that is an intriguing blend of the computational approach to the mind and the traditional psychological continuity view of personhood.[14] If plausible, patternism would explain how one can survive such radical enhancements as those depicted in our thought experiments. Furthermore, it would be an important contribution to the age-old philosophical debate over the nature of persons. So, is it correct? And is patternism even compatible with the radical enhancements fusion-optimists envision? In the next chapter, we'll consider these issues.

CHAPTER SIX

GETTING A MINDSCAN

I teach you the Overman! Mankind is something to be overcome. What have you done to overcome mankind?

FREDRICK NIETZSCHE, *THUS SPOKE ZARATHUSTRA*

*You and I, we are just informational patterns, and we can be up-
graded to a new, superior version—human 2.0, if you will. And from
there, as AI developments continue, further versions of us can be
created, until one day, when the science is just right, in the ultimate
Nietzschean act of self-overcoming, we merge with AI.*

Thus spoke our fusion-optimist.

Let's consider whether the fusion-optimists are right by
mulling over a scenario depicted in the science fiction novel
Mindscan by Robert Sawyer. The protagonist Jake Sullivan has an
inoperable brain tumor. Death could strike him at any moment.
Luckily, Immortex has a new cure for aging and illness—a "mindscan." Immortex scientists will upload his brain configuration
into a computer and "transfer" it into an android body that is
designed using his own body as a template. Although imperfect,
the android body has its advantages—once an individual is uploaded, a backup exists that can be downloaded if one has an
accident. And it can be upgraded as new developments emerge.
Jake will be immortal.

Sullivan enthusiastically signs numerous legal agreements.
He is told that, upon uploading, his possessions will be

transferred to the android, who will be the new bearer of his consciousness. Sullivan's original copy, which will die soon anyway, will live out the remainder of his life on "High Eden," an Immortex colony on the moon. Although stripped of his legal identity, the original copy will be comfortable there, socializing with the other originals who are also still confined to biological senescence.

Sawyer then depicts Jake's perspective while lying in the scanning tube:

> I was looking forward to my new existence. Quantity of life didn't matter that much to me—but quality. And to have time—not only years spreading out into the future, but time in each day. Uploads, after all, didn't have to sleep, so not only did we get all those extra years, we got one-third more productive time. The future was at hand. Creating another me. Mindscan.

But then, a few seconds later:

> "All right, Mr. Sullivan, you can come out now." It was Dr. Killian's voice, with its Jamaican lilt.

> My heart sank. No . . .

> "Mr. Sullivan? We've finished the scanning. If you'll press the red button. . . ." It hit me like a ton of bricks, like a tidal wave of blood. No! I should be somewhere else, but I wasn't. . . .

> I reflexively brought up my hands, patting my chest, feeling the softness of it, feeling it raise and fall. Jesus Christ!

> I shook my head. "You just scanned my consciousness, making a duplicate of my mind, right?" My voice was sneering.

"And since I'm aware of things after you finished the scanning, that means I—this version—isn't that copy. The copy doesn't have to worry about becoming a vegetable anymore. It's free. Finally and at last, it's free of everything that's been hanging over my head for the last twenty-seven years. We've diverged now, and the cured me has started down its path. But this me is still doomed."[1]

Sawyer's novel is a *reductio ad absurdum* of the patternist conception of the person. All that patternism says is that as long as person A has the same computational configuration as person B, A and B are the same person. Indeed, Sugiyama, the person selling the mindscan to Jake, had espoused a form of patternism.[2]

But Jake has belatedly realized a problem with that view, which we shall call "the reduplication problem": Only one person can really be Jake Sullivan. According to patternism, both creatures are Jake Sullivan, because they share the very same psychological configuration. But, as Jake learned, although the creature created by the mindscan process may be a person, it is not the very same person as the original Jake. It is just another person with an artificial brain and body configured like the original. Both feel a sense of psychological continuity with the person who went into the scanner, and both may claim to be Jake, but nonetheless they are not the same person, any more than identical twins are.

Hence, having a particular type of pattern cannot be *sufficient* for personal identity. Indeed, the problem is illustrated to epic proportions later in Sawyer's book when numerous copies of Sullivan are made, all believing they are the original! Ethical and legal problems abound.

A WAY OUT?

The patternist has a response to all this, however. As noted, the reduplication problem suggests that sameness of pattern is not sufficient for sameness of person. You are more than just your pattern. But there still seems to be something right about patternism—for, as Kurzweil notes, throughout the course of your life, your cells change continually; it is your organizational pattern that carries on. Unless you have a religious conception of the person and adopt the soul theory, patternism may strike you as inevitable, at least insofar as you believe there is such a thing as a person to begin with.

In light of these observations, perhaps we should react to the reduplication case in the following way: Your pattern is *essential* to yourself despite not being *sufficient for* a complete account of your identity. Perhaps there is an additional essential property which, together with your pattern, yields a complete theory of personal identity.

What could the missing ingredient be? Intuitively, it must be a requirement that serves to rule out mindscans and, more generally, any cases in which the mind is uploaded. For any sort of uploading case will give rise to a reduplication problem, because uploaded minds can in principle be downloaded again and again.

Now think about your own existence in space and time. When you go out to get the mail, you move from one spatial location to the next, tracing a path in space. A spacetime diagram can help us visualize the path one takes throughout one's life. Collapsing the three spatial dimensions into one (the vertical axis) and taking the horizontal axis to signify time, consider the following typical trajectory:

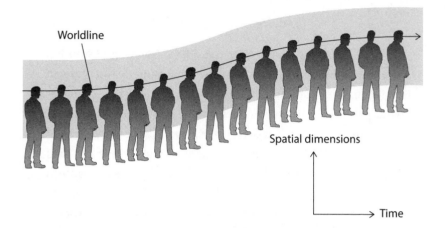

Notice that the figure carved out looks like a worm; you, like all physical objects, carve out a sort of "spacetime worm" over the course of your existence.

This, at least, is the kind of path that ordinary people—those who are neither posthumans nor superintelligences—carve out. But now consider what happened during the mindscan. Again, according to patternism, there would be two copies of the very same person. The copy's spacetime diagram would look like the one pictured on the following page.

This is bizarre. It appears that Jake Sullivan exists for 42 years, has a scan, and then somehow instantaneously moves to a different location in space and lives out the rest of his life. This is radically unlike normal survival. This alerts us that something is wrong with pure patternism: It lacks a requirement for spatiotemporal continuity.

This additional requirement would seem to solve the reduplication problem. On the day of the mindscan, Jake went into the laboratory and had a scan; then he left the laboratory and went directly into a spaceship and flew to exile on the moon. It is this man—the one who traces a continuous trajectory

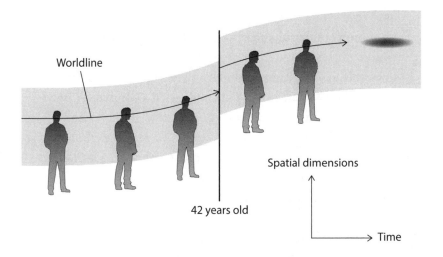

Worldline

Spatial dimensions

42 years old

Time

through space and time—who is the true Jake Sullivan. The android is an unwitting impostor.

This response to the reduplication problem only goes so far, however. Consider Sugiyama, who, when selling his mindscan product, ventured a patternist pitch. If he had espoused pattern-ism together with a spatiotemporal continuity clause, he would have to admit that his customers would not become immortal, and few would have signed up for the scan. That extra ingredi-ent would rule out a mindscan (or any kind of uploading, for that matter) as a means to ensure survival. Only those wishing to have a replacement for themselves would sign up.

There is a general lesson here for the transhumanist or fusion-optimist: If one opts for patternism, enhancements like uploading to avoid death or to facilitate further enhancements are not really "enhancements;" they can even result in death. *The fusion-optimist should sober up and not offer such procedures as enhancements.* When it comes to enhancement, there are in principle limitations to what technology can deliver. Making copies of a mind does not count as an enhancement, because each individual mind still carries on, and it is subject to the limitations of its substrate. (Ironically, the proponent of the soul theory is in better shape here, because perhaps the soul does upload. If so, Jake could well wake up to find himself in the android body, while his original flesh is left a zombie, stripped of conscious experience. Who knows?)

Now let's pause and take a breath. We've accomplished a lot in this chapter. We began by thinking about the *Mindscan* case, and a "reduplication problem" arose for patternism. This led us to discard the original form of patternism as false. I then suggested a way to modify patternism to arrive at a more realistic position. This meant adding a new element to the view—namely, the spatiotemporal continuity clause—which required that there be spatiotemporal continuity of a pattern for it to survive. I called this *modified patternism*. Modified patternism may strike you as more sensible, but notice that it doesn't serve the fusion-optimist well, because it means that uploading is incompatible with survival, since uploading violates the spatiotemporal-continuity requirement.

But what about other AI-based enhancements? Are these ruled out as well? Consider, for instance, selecting a bundle of enhancements at the Center for Mind Design. These enhancements could dramatically alter your mental life, yet they do not involve uploading, and it is not obvious that spatiotemporal continuity would be violated.

Indeed, the fusion-optimist could point out that one could still merge with AI through a series of gradual but cumulatively significant enhancements that added AI-based components inside the head, slowly replacing neural tissue. This wouldn't be uploading, because one's thinking would still be inside the head, but the series still amounts to an attempt to transfer one's mental life to another substrate. When the series was completed, if it worked, the individual's mental life would have migrated from a biological substrate to a nonbiological one, such as silicon. And the fusion-optimist would be right: Humans can merge with AI.

Would it work? Here we need reconsider some issues raised in Chapter Five.

DEATH OR PERSONAL GROWTH?

Suppose you are at the Center for Mind Design, and you are gazing at the menu, considering buying a certain bundle of enhancements. Longing to upgrade yourself, you reluctantly consider whether modified patternism might be true. And you wonder: If I am a modified pattern, what happens to me when I add the enhancement bundle? My pattern will surely change, so would I die?

To determine whether this would be the case, the modified patternist would need to give a more precise account of what a "pattern" is and when different enhancements do and do not constitute a deadly break in the pattern. The extreme cases seem clear—for instance, as discussed, mindscans and duplication are ruled out by the spatiotemporal-continuity requirement. And, furthermore, because both versions of patternism are closely related to the older psychological-continuity approach, the modified patternist will likely want to say that a memory erasure process that erased several difficult years from one's childhood is an unacceptable alteration of one's pattern, removing too many of one's memories and changing the person's nature. In contrast, mere everyday cellular maintenance by nanobots swimming through your bloodstream to overcome the slow effects of aging would, presumably, not affect the identity of the person, for it wouldn't alter one's memories.

The problem is that the middle-range cases are unclear. Maybe deleting a few bad chess-playing habits is kosher, but what about more serious mindsculpting endeavors, like the enhancement bundle you were considering, or even adding a single cognitive capacity, for that matter? Or what about raising your IQ by 20 points or erasing all your memory of some personal relationship, as in the film *Eternal Sunshine of the Spotless Mind*? The path to superintelligence may very well be at the end of a gradual path through a series of these sorts of enhancements. But where do we draw the line?

Each of these enhancements is far less radical than uploading, but each one could be an alteration in the pattern that is incompatible with the preservation of the original person. And cumulatively, their impact on one's pattern can be quite significant. So again, what is needed is a clear conception of what a pattern is, what changes in pattern are acceptable, and why. Without a firm handle on this issue, the transhumanist developmental trajectory is, for all we know, a technophile's alluring path to suicide.

This problem looks hard to solve in a way that is compatible with preserving the very idea that we can be identical over time to some previous or future self. Determining a boundary point threatens to be an arbitrary exercise. Once a boundary is selected, an example can be provided, suggesting the boundary should be pushed outward, ad nauseam. But appreciate this point too long, and it may lead to a dark place: If one finds patternism or modified patternism compelling to begin with, how is it that one's pattern truly persists over time, from the point of infancy until maturity, during which time there are often major changes in one's memories, personality, and so on? Why is there a persisting self at all?

Indeed, even a series of gradual changes cumulatively amounts to an individual, B, who is greatly altered from her childhood self, A. Why is there really a relation of identity that holds between A and B, instead of an ancestral relation: A's being the ancestor of B? Put differently, how can we tell if that future being who exists, after all these enhancements, is really us, and not, instead, a different person—a sort of "descendent" of ours?

It is worth pausing for a moment to reflect on the relationship between the ancestor and the descendant, although it is a bit of an aside. Suppose you are the ancestor. Your connection to your descendent resembles a parent-child relationship, but in some ways, it is more intimate, because you have first-person knowledge of the past of this new being. He or she is your mindchild. You have literally lived those past moments. In contrast, although we may feel closely connected to our children's lives, we do not literally see the world through their eyes. In another sense, the relationship is weaker than many parent-child connections, however. Unless you are some sort of time traveler, the two of you will never even be in the same room. Like a woman who dies in childbirth, you will never meet the descendant who follows.

Perhaps your mindchild will come to mourn your passing, regarding you with deep affection and appreciating that the end of your life was the beginning of their own. For your part, you may feel a special connection to the range of experiences that will be enabled by the enhancements that you are about to purchase for your mindchild's benefit. You may even feel a special connection to a being who you know will be nothing like you. For instance, in an act of benevolent mindsculpting, perhaps you would deliberately set out to create a superintelligent being, knowing that if you succeed, you will die.

In any case, the main point of this section has been to show you that even modified patternism faces a key challenge—it needs to tell us when a shift in one's pattern is compatible with survival, and when it is not. Until it does, a cloud will hang over the fusion-optimist project. And this is not the only challenge for modified patternism.

DITCHING YOUR SUBSTRATE?

Modified patternism faces a second problem as well, one that challenges the very possibility that an individual could transfer to a different substrate, even with no cognitive or perceptual enhancements.

Suppose that it is 2050, and people are getting gradual neural regeneration procedures as they sleep. During their nightly slumbers, nanobots slowly import nanoscale materials that are computationally identical to the original materials. The nanobots then gradually remove the old materials, setting them in a small container beside the person's bed.

By itself, this process is unproblematic for modified patternism. But now suppose there is an optional upgrade to the regeneration service for those who would like to make a backup copy of their brains. If one opts for this procedure, then during the nightly process, the nanobots take the replaced materials out of the dish and place them inside a cryogenically frozen biological brain. Suppose that by the end of the process, the materials in the frozen brain have been entirely replaced by the person's original neurons, and they are configured the same way that they were in the original brain.

Now, suppose you choose to undergo this add-on procedure alongside your nightly regeneration. Over time, this second brain comes to be composed of the very same material as your brain originally was, configured in precisely the same manner. Which one is you? The original brain, which now has entirely different neurons, or the one with all your original neurons? The modified patternist has this to say about the neural regeneration case: You are the creature with the brain with entirely different matter, as this creature traces a continuous path through space-time. But now, things go awry: Why is spatiotemporal continuity supposed to outweigh other factors, like being composed of the original material substrate?

Here, to be blunt, my intuitions crap out. I do not know whether this thought experiment is technologically feasible, but nevertheless, it raises an important conceptual flaw. We are trying to find out what the essential nature of a person is, so we'd like to find a solid justification for selecting one option over the other. Which is the deciding factor that is supposed to enable one to survive, in a case in which psychological continuity holds—being made of one's original parts or preserving spatiotemporal continuity?

These problems suggest that modified patternism needs a good deal of spelling out. And remember, it wasn't consistent with uploading in any case. The original patternist view held that one can survive uploading, but we discarded the view as deeply problematic. Until the fusion-optimist provides a solid justification for her position, it is best to view the idea of merging with AI with a good deal of skepticism. Indeed, after considering the vexing issue of persistence, perhaps we should even question the wisdom of a limited integration with AI, as middle-range enhancements are not clearly compatible with survival. Furthermore, even enhancements that merely involve the rapid or even gradual replacement of parts of one's brain, without even enhancing one's cognitive or perceptual skills, may be risky.

METAPHYSICAL HUMILITY

At the outset of the book, I asked you to imagine a shopping trip at the Center for Mind Design. You can now see how deceptively simple this thought experiment was. Perhaps the best response to the ongoing controversy over the nature of persons is to take a simple stance of *metaphysical humility*. Claims about survival that involve one "transferring" one's mind to a new type of substrate or making drastic alterations to one's brain should be carefully scrutinized. As alluring as greatly enhanced intelligence or digital immortality may be, we've seen that there is much disagreement in the personal-identity literature over whether any of these "enhancements" would extend life or terminate it.

A stance of metaphysical humility says that the way forward is public dialogue, informed by metaphysical theorizing. This may sound like a sort of intellectual copout, suggesting that scholars are of little use on this matter. But I am not suggesting that further metaphysical theorizing is useless; on the contrary, my hope is that this book illustrates the life-and-death import of further metaphysical reflection on these issues. The point is that ordinary individuals must be capable of making informed decisions about enhancement, and if the success of an

enhancement rests on classic philosophical issues that are difficult to solve, the public needs to appreciate this problem. A plu-
ralistic society should recognize the diversity of different views
on these matters and not assume that science alone can answer
questions about whether radical forms of brain enhancement
are compatible with survival.

All this suggests that one should take the transhumanist approach to radical enhancement with a grain of salt. As "The
Transhumanist Frequently Asked Questions" indicates, the
development of enhancements like brain uploading or adding brain chips to augment intelligence or radically alter one's
perceptual abilities are key enhancements invoked by the
transhumanist view of the development of the person.[3] Such
enhancements sound strangely like the thought experiments
philosophers have used for years as problem cases for various
theories of the nature of persons, so it shouldn't surprise us
one bit that the enhancements aren't as attractive as they might
seem at first.

We've learned that the *Mindscan* example suggests that one
should not upload (at least not if one hopes to survive the
process) and that the patternist needs to modify her theory
to rule out such cases. Even with this modification in hand,
however, transhumanism and fusion-optimism still require a
detailed account of what constitutes a break in a pattern versus a mere continuation of it. Without progress on this issue,
it will not be clear if medium-range enhancements, such as
adding neural circuitry to make oneself smarter, are safe. Finally, the nanobot case warns against migrating to a different
substrate, even if one's mental abilities remain unchanged.
Given all this, it is fair to say that the fusion-optimists or transhumanists have failed to support their case for enhancement.

Indeed, "The Transhumanist Frequently Asked Questions" notes that transhumanists are keenly aware that this issue has been neglected:

> While the concept of a soul is not used much in a naturalistic philosophy such as transhumanism, many transhumanists do take an interest in the related problems concerning personal identity (Parfit 1984) and consciousness (Churchland 1988). These problems are being intensely studied by contemporary analytic philosophers, and although some progress has been made, e.g. in Derek Parfit's work on personal identity, they have still not been resolved to general satisfaction.[4]

Our discussion also raises general lessons for all parties involved in the enhancement debate, even where purely biological enhancements are concerned. When one considers the enhancement debate through the lens of the metaphysics of personhood, new dimensions of the debate unfurl. The literature on the nature of persons is extraordinarily rich, and when one defends or rejects a given enhancement, it is important to determine whether one's stance on the enhancement in question is truly supported by (or even compatible with) leading views on the nature of persons.

Perhaps, alternately, you grow weary of all this metaphysics. You may suspect that we have to fall back on social conventions concerning what we commonly consider to be persons, because metaphysical theorizing will never conclusively resolve what persons are. However, not all conventions are worthy of acceptance, so one needs a manner of determining which conventions should play an important role in the enhancement debate and which ones should not. And it is hard to accomplish this without getting clear on one's conception of persons. Furthermore, it is difficult to avoid at least implicitly relying on a conception of persons when reflecting on the case for and against enhancement. What is it that ultimately grounds your decision to enhance or not to enhance, if not that it will somehow improve you? Are you perhaps merely planning for the well-being of your successor?

We will return to personal identity in Chapter Eight. There, we will consider a related position on the fundamental nature of mind that says that the mind is software. But let's pause for a moment. I'd like to raise the ante a bit. We've seen that each of us alive today may be one of the last biological rungs on the evolutionary ladder that leads from the first living cell to synthetic intelligence. On Earth, *Homo sapiens* may not be the most intelligent species for that much longer. In Chapter Seven, I'd like to explore the evolution of mind in a cosmic context. The minds on Earth—past, present, and future—may be but a small speck in the larger space of minds, a space that spans all of spacetime. As I write this, civilizations elsewhere in universe may be having their own singularities.

A UNIVERSE OF SINGULARITIES

In your mind's eye, zoom away from Earth. Envision Earth becoming but a "pale blue dot" in outer space, to use an expression of Carl Sagan's. Now zoom out of the Milky Way Galaxy. The scale of the universe is truly staggering. We are but one planet in an immense, expanding universe. Astronomers have already discovered thousands of exoplanets, planets beyond our solar system, many of which are Earthlike—they seem to have the sort of conditions that led to the development of life on Earth. As we gaze up into the night sky, life could be all around us.

This chapter will illustrate that the technological developments we are witnessing today on Earth may have all happened before, elsewhere in the universe. That is, the universe's greatest intelligences may be synthetic, having grown out of civilizations that were once biological.[1] The transition from biological intelligence to synthetic intelligence may be a general pattern, instantiated over and over, throughout the cosmos. If a civilization develops the requisite AI technology, and cultural conditions are favorable, the transition from biological to postbiological may take only a few hundred years. As you read these words, there may be thousands, or even millions, of other worlds that have developed AI technology.

In reflecting on postbiological intelligence, we are not just considering the possibility of alien intelligence—we may also be reflecting on the nature of ourselves or our descendants as well, for we've seen that human intelligence may itself become postbiological. So, in essence, the line between "us" and "them" blurs, as our focus moves away from biology to the difficult task of understanding the computations and behaviors of superintelligence.

Before we delve further into this, a note on the expression "postbiological." Consider a biological mind that achieves superintelligence through purely biological enhancements, such as nanotechnologically enhanced neural minicolumns. This creature would be postbiological, although many wouldn't call it an "AI." Or consider a computronium that is built out of purely biological materials, like the Cylon Raider in the reimagined *Battlestar Galactica* TV series. The Cylon Raider is artificial, and postbiological.

The key point is that there is no reason to expect humans to be the highest form of intelligence out there. It is humbling to conceive of this, but we may be intellectual lightweights when viewed on a galactic scale, at least until we enhance our minds in radical ways. The intellectual gap between an unenhanced human and alien superintelligence could be like that between us and goldfish.

THE POSTBIOLOGICAL COSMOS

In the field of astrobiology, this position has been called the *postbiological cosmos* approach. This approach says that the members of the most intelligent alien civilizations will be superintelligent AIs. What is the rationale for this? Three observations, when considered together, motivate this conclusion.

1. It takes only a few hundred years—a cosmic eyeblink—
   for a civilization to go from pre-industrial to
   postbiological.

Many have urged that once a society creates the technology
that could put them in touch with intelligent life on other planets, there is only a short window before they change their own
paradigm from biology to AI—perhaps only a few hundred
years.[2] This makes it more likely that the aliens we encounter,
if we encounter any, would be postbiological. Indeed, the short-
window observation seems to be supported by human cultural
evolution, at least thus far. Our first radio signals occurred only
about 120 years ago, and space exploration is only about 50 years
old, but many Earthlings are already immersed in digital technology, such as smartphones and laptop computers. Currently,
billions of dollars are being poured into the development of
sophisticated AI, which is now expected to change the face of
society within the next several decades.

A critic may object that this line of thinking employs "$N=1$
reasoning." Recall that this is a form of reasoning that mistakenly generalizes from the human case to the case of alien life.
But it strikes me as being unwise to discount arguments based
on the human case—human civilization is the only one we
know of, and we had better learn from it. It is no great leap to
claim that other technological civilizations will develop technologies to advance their intelligence and gain an adaptive advantage. And we've seen that synthetic intelligence will likely be
able to radically outperform the unenhanced brain.

An additional objection to my short-window observation
points out that nothing I have said thus far suggests that humans
will be superintelligent; I have said just that future humans
will be postbiological. But postbiological beings may not be so

advanced as to be superintelligent. So even if one is comfortable reasoning from the human case, the human case does not actually support the claim that the members of the most advanced alien civilizations will be superintelligent.

This is a valid objection, but I think the other considerations that follow show that alien intelligence is also likely to be superintelligent.

2. Alien civilizations may have already been around for billions of years.

Proponents of SETI ("the Search for Extraterrestrial Intelligence") have often concluded that alien civilizations would be much older than our own, if they exist. As the former NASA chief historian, Steven Dick, observes: "all lines of evidence converge on the conclusion that the maximum age of extraterrestrial intelligence would be billions of years, specifically [it] ranges from 1.7 billion to 8 billion years."[3] This is not to say that all life evolves into intelligent, technological civilizations. It is just to say that there are much older planets than Earth. Insofar as intelligent, technological life does evolve on even some of them, these alien civilizations are projected to be millions or billions of years older than us, so many could be vastly more intelligent than we are. They would be superintelligent, by our standards. It is humbling to conceive of this, but we may be galactic babies. When viewed on a cosmic scale, Earth is but a playpen for intelligence.

But would the members of these superintelligent civilizations be forms of AI? Even if they were biological and had received brain enhancements, their superintelligence would be reached by artificial means, which leads me to my third observation:

3. It is likely that these synthetic beings will not be biologically based.

I've already observed that silicon appears to be a better medium for information processing than the brain itself. In addition, other, superior kinds of microchips are currently under development, such as those based on graphene and carbon nanotubes. The number of neurons in the human brain is limited by cranial volume and metabolism, but computers can be remotely connected across the globe. AIs can be constructed by reverse-engineering the brain and improving on its algorithms. And AI is more durable and can be backed up.

There is one thing that would stand in the way: the very worries I have been expressing in this book. Like human philosophers, alien thinkers may also come to appreciate the difficult and possibly intractable issues of personal identity raised by cognitive enhancements. Maybe they resisted the pull of radical enhancement, as I have been urging us to do.

Unfortunately, I think there is a good chance that some civilizations succumbed. This doesn't necessarily mean that the members of these civilizations become zombies; hopefully, the superintelligences are conscious beings. But it does mean that members who "enhanced" may have died. Perhaps these civilizations didn't halt enhancements, because they mistakenly believed they found clever solutions to the philosophical puzzles. Or perhaps on some worlds, the aliens weren't even philosophical enough to reflect on these issues. And perhaps on some other distant worlds they did, but they concluded, based on reflections of alien philosophers who have views akin to the Buddha or Parfit, that there is no real survival anyway. Not believing in the self at all, they opted to upload. They might have been what the philosopher Pete Mandik called "metaphysically daring:" willing to make a leap of faith that consciousness or the self can be preserved when one transfers the informational structure of the brain from tissue to silicon chips.[4] Another possibility is that certain alien civilizations take great care in enhancing an individual during their lifetime, so as to not violate certain principles of personal identity, but they use reproductive technologies to create new members of the species with highly enhanced abilities. Other civilizations may have simply lost control of their AI creations and been unwittingly supplanted.

Whichever intelligent civilizations didn't halt enhancement efforts, for whatever reason, became the most intelligent civilizations in the universe. Whether these aliens are good philosophers or not, their civilizations still reap the intellectual benefits. As Mandik suggests, systems that have high degrees of metaphysical daring could, through making many more digital backups of themselves, be more fit in a Darwinian sense than more cautious beings in other civilizations.[5]

Furthermore, I've noted that AIs are more likely to endure space travel, being both more durable and capable of backup, so they will likely be the ones to colonize the universe, if anyone does. They may be the kind of the creatures we Earthlings first encounter, even if they aren't the most common.

In sum, there seems to be a short window of time from the development of space travel and communications technology to the development of postbiological minds. Extraterrestrial civilizations will have passed through this window long ago. They are likely to be vastly older than us, and thus they would have already reached not just a postbiological stage, but superintelligence. Finally, at least some will be AIs rather than biological creatures, because silicon and other materials are a superior medium for information processing. From all this, I conclude that if life is indeed present on many other planets, and if advanced civilizations do tend to develop and then survive, the members of most advanced alien civilizations will likely be superintelligent AIs.

The science fiction–like flavor of these issues can encourage misunderstanding, so it is worth stressing that I am not claiming that most life in the universe is nonbiological. Most life on Earth itself is microbial. Nor am I saying that the universe will be "controlled" or "dominated" by a single superintelligent AI, akin to *Skynet* from the *Terminator* films, although it is worth reflecting on AI safety in the context of these issues. (Indeed, I shall do so shortly.) I am merely suggesting that the members of the most advanced alien civilizations will be superintelligent AIs.

Suppose I am right. What should we make of this? Here, current debates over AI on Earth are telling. Two important issues—the so-called control problem and the nature of mind and consciousness—impact our understanding of what superintelligent alien civilizations may be like. Let's begin with the control problem.

THE CONTROL PROBLEM

Advocates of the postbiological cosmos approach suspect that machines will be the next phase in the evolution of intelligence. You and I, how we live and experience life right now, are just an intermediate step to AI, a rung on the evolutionary ladder. These individuals tend to have an optimistic view of the postbiological phase of evolution. Others, in contrast, are deeply concerned that humans could lose control of superintelligence, because a superintelligence could rewrite its own code and outthink any safeguards we build in. AI could be our greatest invention and our last one. This has been called the "control problem"—how we Earthlings can control an AI that is both inscrutable and vastly smarter than us.

We've seen that superintelligent AI could be developed during a technological singularity, a point at which ever-more-rapid technological advances—especially an intelligence explosion—reach a point at which humans can no longer predict or understand the technological changes as they unfold. But even if superintelligent AI arises in a less dramatic fashion, there may be no way for us to foresee or control the goals of AI. Even if we could decide on what moral principles to build into our machines, moral programming is difficult to specify in a foolproof way, and any such programming could be rewritten by a superintelligence in any case. A clever machine could bypass safeguards, such as kill switches, and could potentially pose an existential threat to biological life.

The control problem is a serious problem—perhaps it is even insurmountable. Indeed, upon reading Bostrom's compelling book on the control problem, *Superintelligence: Paths, Dangers and Strategies,*[6] scientists and business leaders such as Stephen Hawking and Bill Gates were widely reported by the world media as commenting that superintelligent AI could threaten the human race. At this time, millions of dollars are pouring into organizations devoted to AI safety and some of the finest minds in computer science are working on the problem. Let us consider the implications of the control problem for the SETI project.

ACTIVE SETI

The usual approach to search for life in the universe is to listen for radio signals from extraterrestrial intelligence. But some astrobiologists think we should go a step further. Advocates of Active SETI hold that we should also be using our most

Satellite. Courtesy of the Arecibo Observatory, a facility of the NSF

powerful radio transmitters, such as the giant dish-telescope at Arecibo, Puerto Rico, (pictured above) to send messages in the direction of the stars that are nearest to Earth in order to initiate a conversation.[7]

Active SETI strikes me as reckless when one considers the control problem, however. Although a truly advanced civilization would probably have no interest in us, we should not call attention to ourselves, as an encounter with even one hostile civilization among millions could be catastrophic. Maybe one day we will reach the point at which we can be confident that alien superintelligences do not pose a threat to us, but we have no justification for such confidence just yet. Proponents of Active SETI argue that a deliberate broadcast would not make us any more vulnerable than we already are, pointing out that our radar and radio signals are already detectable. But these signals are fairly weak and quickly blend with natural galactic noise. We would be playing with fire if we transmitted stronger signals that were intended to be heard.

The safest mindset is intellectual humility. Indeed, barring blaringly obvious scenarios in which alien ships hover over Earth, as in films like *Arrival* and *Independence Day*, I wonder if we could even recognize the technological markers of a truly advanced superintelligence. Some scientists project that superintelligent AIs could be found near black holes, feeding off their energy.[8] Alternately, perhaps superintelligences would create Dyson spheres, megastructures such as that pictured on the following page, which harness the energy of an entire star.

But these are just speculations from the vantage point of our current technology; it's simply the height of hubris to claim that we can foresee the computational structure or energy needs of a civilization that is millions or even billions of years ahead of our own. For what it's worth, I suspect that we will not detect or be contacted by alien superintelligences until our own civilization becomes superintelligent. It takes one to know one.

Although many superintelligences would be beyond our grasp, perhaps we can be more confident when speculating on the nature of "early" superintelligences—that is, those that emerge from a civilization that was previously right on the cusp

Dyson sphere

of developing superintelligence. Some of the first superintelligent AIs could have cognitive systems that are modeled after biological brains—the way, for instance, that deep-learning systems are roughly modeled on the brain's neural networks. So their computational structure might be comprehensible to us, at least in rough outlines. They may even retain goals that biological beings have, such as reproduction and survival. I will turn to this issue of early superintelligence in more detail shortly.[9]

But superintelligent AIs, being self-improving, could quickly transition to an unrecognizable form. Perhaps some superintelligences will opt to retain cognitive features that are similar to those of the species they were originally modeled after, placing a design ceiling on their own cognitive architecture. Who knows? But without a ceiling, an alien superintelligence could quickly outpace our ability to make sense of its actions or even look for it.

An advocate of Active SETI will point out that this is precisely why we should send signals into space—let the superintelligent civilizations locate us, and let them design means of contact they judge to be tangible to an intellectually inferior species like us. While I agree this is reason to consider Active SETI, I believe that the possibility of encountering a dangerous superintelligence outweighs it. For all we know, malicious superintelligences could infect planetary AI systems with viruses, and wise civilizations build cloaking devices; perhaps this is why we haven't yet detected anyone. We humans may need to reach our own singularity before embarking on Active SETI. Our own superintelligent AIs will be able to inform us of the prospects for galactic AI safety and how we should go about recognizing signs of superintelligence elsewhere in the universe. Again, "it takes one to know one" is the operative slogan.

SUPERINTELLIGENT MINDS

The postbiological cosmos approach involves a radical shift in our usual perspective about intelligent life in the universe. Normally, we expect that if we encountered advanced alien intelligence, we would encounter creatures with very different *biological* features than us, but that much of our intuition about minds would still apply. But the postbiological cosmos approach suggests otherwise.

In particular, the standard view is that if we ever encountered advanced alien creatures, they would still have minds like ours in an important sense—there would be something it is like, from the inside, to be them. We've seen that throughout your daily life, and even when you dream, it feels like something to be you. Likewise, there is also something that it is like to be a biological alien, if such exist—or so we tend to assume. But would a superintelligent AI even have conscious experience? If it did, could we tell? And how would its inner life, or lack thereof, impact its capacity for empathy and the kind of goals it has? Raw intelligence is not the only issue to consider when thinking about contact with extraterrestrials.

We considered these issues in detail in earlier chapters, and we can now appreciate their cosmic import. I've noted that the question of whether an AI could have an inner life should be key to how we value its existence, because consciousness is central to our judgment of whether it is a self or person. An AI could even be superintelligent, outperforming humans in every cognitive and perceptual domain, but if it doesn't feel like anything to be the AI, it difficult to view these beings as having the same value as conscious beings, being selves or persons. And conversely, I've observed that whether AI is conscious may also be key to how it values *us*: A conscious AI could recognize in us the capacity for conscious experience.

Clearly, the issue of machine consciousness could be central to how humans would react to the discovery of superintelligent aliens. One way that humanity will process the implications of contact will be through religion. And although I hesitate to speak for world religions, discussions with my colleagues working in astrobiology at the Center of Theological Inquiry, Princeton, suggest that many would reject the possibility that AIs could have souls or are somehow made in God's image, if they are not even conscious beings. Indeed, Pope Francis has recently commented that he would baptize an extraterrestrial.[10] But I wonder how Pope Francis would react if asked to baptize an AI, let alone one that is not capable of consciousness.

This isn't just a romantic question of whether ETs will enjoy sunsets or possess souls, but an existential one for us. Because even if the universe were stocked full of AIs of unbelievable intelligence, why would those machines place any value on conscious biological intelligences? Nonconscious machines cannot experience the world and, lacking that awareness, may be incapable of genuine empathy or even intellectual concern for outmoded creatures.

BIOLOGICALLY INSPIRED SUPERINTELLIGENCES

Thus far, I've said little about the structure of superintelligent alien minds. And little is all we can say: Superintelligence is by definition a kind of intelligence that outthinks humans in every domain. In an important sense, we cannot predict or fully understand how it will think. Still, we may be able to identify a few important characteristics, at least in broad strokes.

Nick Bostrom's recent book on superintelligence focuses on the development of superintelligence on Earth, but we can draw from his thoughtful discussion. Bostrom distinguishes three kinds of superintelligence:

Speed superintelligence: a superintelligence having rapid-fire cognitive and perceptual abilities. For instance, even a human emulation or upload could in principle run so fast that it could write a PhD thesis in an hour.

Collective superintelligence: the individual units need not be superintelligent, but the collective performance of the individual members vastly outstrips the intelligence of any individual human.

Quality superintelligence: an intelligence that computes at least as fast as humans think and that also outthinks humans in every domain.[11]

Bostrom indicates that any of these kinds of superintelligence could exist alongside one or more of the others.

An important question is whether we can identify common goals that these types of superintelligences could share. Bostrom's suggests the following thesis:

The Orthogonality Thesis: Intelligence and final goals are orthogonal—"more or less any level of intelligence could in principle be combined with more or less any final goal."[12]

Put simply, just because an AI is smart doesn't mean it has perspective; all the intelligence of a superintelligent being could be marshaled to absurd ends. (This reminds me a bit of academic politics, in which so much intelligence can be utterly wasted on petty or even perverse goals). Bostrom is careful to underscore that a great many unthinkable kinds of superintelligences could be developed. At one point in the book, he raises a sobering example of a superintelligence that runs a paper-clip factory. Its final goal is the banal task of manufacturing paper clips.[13] Although this may initially strike you as harmless endeavor (but hardly a life worth living), Bostrom's sobering point is that superintelligence could utilize every form of matter on Earth in support of this goal, wiping out biological life in the process.

The paper-clip example illustrates that superintelligence could be of an unpredictable nature, having thinking that is "extremely alien" to us.[14] Although the final goals of superintelligence are difficult to predict, Bostrom singles out several instrumental goals as being likely, given that they support any final goal whatsoever:

The Instrumental Convergence Thesis: "Several instrumental values can be identified which are convergent in the sense that their attainment would increase the chances of the agent's goal being realized for a wide range of final goals and a wide range of situations, implying that these instrumental values are likely to be pursued by a broad spectrum of situated intelligent agents."[15]

The goals that Bostrom identifies are resource acquisition, technological perfection, cognitive enhancement, self-preservation, and goal-content integrity (i.e., that a superintelligent being's future self will pursue and attain those same goals). He underscores that self-preservation can involve group or individual preservation, and that it may play second fiddle to the preservation of the species the AI was designed to serve.

Bostrom does not speculate about superintelligent alien minds in his book, but his discussion is suggestive. Let us call an alien superintelligence that is based on reverse engineering an alien brain, including uploading it, a "biologically inspired superintelligent alien" (BISA). Although BISAs are inspired by the brains of the original species that the superintelligence is derived from, their algorithms may depart from those of their biological model at any point.

BISAs are of particular interest in the context of alien superintelligence, because they form a special class in the full spectrum of possible AIs. If Bostrom is correct that there are many ways superintelligence can be built, superintelligent AIs will be highly heterogeneous, with members generally bearing little resemblance to one another. It may turn out that of all superintelligent AIs, BISAs bear the most resemblance to one another by virtue of their biological origins. In other words, BISAs may be the most cohesive subgroup, because the other members are so different from one another. BISAs may be the single most common form of alien superintelligence out there.

You may suspect that because BISAs could be scattered across the galaxy and generated by multitudes of species, there is little interesting that we can say about the class of BISAs. You may object that it is useless to theorize about BISAs, as they can change their basic architecture in numerous, unforeseen ways, and any biologically inspired motivations can be constrained by programming. But notice that BISAs have two features that may give rise to common cognitive capacities and goals:

1. BISAs are descended from creatures that had motivations like: find food, avoid injury and predators, reproduce, cooperate, compete, and so on.
2. The life forms that BISAs are modeled on have evolved to deal with biological constraints like slow processing speed and the spatial limitations of embodiment.

Could these features yield traits common to members of many superintelligent alien civilizations? I suspect so.

Consider feature 1. Intelligent biological life tends to be primarily concerned with its own survival and reproduction, so it is more likely that a BISA would have final goals involving its own survival and reproduction, or at least the survival and reproduction of the members of its society. If BISAs are interested in reproduction, we might expect that, given the massive amounts of computational resources at their disposal, BISAs would create simulated universes stocked with artificial life and even intelligence or superintelligence. If these creatures were intended to be "mindchildren," they may retain the goals listed in feature 1 as well.

Likewise, if a superintelligence continues to take its own survival as a primary goal, it may not wish to change its architecture fundamentally. It may opt for a series of smaller improvements that nevertheless gradually lead the individual toward super-intelligence. Perhaps, after reflecting on the personal-identity debate, BISAs tend to appreciate the vexing nature of the issues, and they think: "Perhaps, when I fundamentally alter my architecture, I will no longer be me." Even a being that is an upload, and which believes that it is not identical to the creature that uploaded, may nevertheless wish not to alter the traits that were most important to their biological counterparts during their biological existence. Remember, uploads are isomorphs (at least at the time they are uploaded), so these are traits that they identify with, at least initially. Superintelligences that reason in this way may elect to retain biological traits.

Consider feature 2. Although I have noted that a BISA may not wish to alter its architecture fundamentally, it or its designers may still move away from the original biological model in all sorts of unforeseen ways. Even then, though, we could look for cognitive capacities that are useful to keep: cognitive capacities that sophisticated forms of biological intelligence are likely to have and that enable the superintelligence to carry out its final and instrumental goals. We could also look for traits that are not likely to be engineered out, as they do not detract the BISA from its goals. We might expect the following, for instance.

1. *Learning about the computational structure of the brain of the species that created the BISA can provide insight into the BISA's thinking patterns.* One influential means of understanding the computational structure of the brain in cognitive science is through the field of connectomics, a field that aims to provide a connectivity map or wiring diagram of the brain, called the "connectome."[16]

Although it is likely that a given BISA will not have the same kind of connectome as the members of the original species did, some of the functional and structural connections may be retained, and interesting departures from the originals may be found. So, this may sound right out of *The X-files*, but an alien autopsy could be quite informative!

2. *BISAs may have viewpoint-invariant representations.* Consider walking up to your front door. You've walked this path hundreds, maybe thousands of times, but technically, you see things from slightly different angles each time, as you are never positioned in exactly the same way twice. But obviously, the path is a familiar one, and this is because at a high level of processing, your brain has internal representations of the people and objects that you interact with that do not vary with your angle or position with respect to them. For instance, you have an abstract notion of door that is independent of the precise appearance of any given door.

Indeed, it strikes me as difficult for biologically based intelligences to evolve without such representations, as they enable categorization and prediction.[17] Invariant representations arise because a system that is mobile needs a means of identifying items in its ever-changing environment, so we would expect biologically based systems to have them. A BISA would have little reason to give up invariant representations insofar as it remains mobile or has mobile devices sending it information remotely.

3. *BISAs will have language-like mental representations that are recursive and combinatorial.* Notice that human thought has the crucial and pervasive feature of being combinatorial. Consider the thought that wine is better in Italy than in China. You may have never had this

thought before, but you were able to understand it. The key is that thoughts are built out of familiar constituents and combined according to rules. The rules apply to constructions out of primitive constituents, which are themselves constructed grammatically. Grammatical mental operations are incredibly useful: It is the combinatorial nature of thought that allows one to understand and produce these sentences on the basis of one's antecedent knowledge of the grammar and atomic constituents (e.g., wine, China). Relatedly, thought is productive: In principle, one can entertain and produce an infinite number of distinct representations, because the mind has a combinatorial syntax.[18]

Brains need combinatorial representations, because there are infinitely many possible linguistic representations, and the brain only has a finite storage space. Even a superintelligent system would benefit from combinatorial representations. Although a superintelligent system could have computational resources that are so vast that it can store every possible utterance or inscription sentence, it would be unlikely that it would trade away such a marvelous innovation of biological brains. If it did, it would be less efficient, because there is the potential of a sentence not being in its storage, which must be finite.

4. *BISAs may have one or more global workspaces.* When you search for a fact or concentrate on something, your brain grants that sensory or cognitive content access to a "global workspace" where the information is broadcast to attentional and working memory systems for more concentrated processing, as well as to the massively parallel channels in the brain.[19] The global workspace

operates as a singular place where important information from the senses is considered in tandem, so that the creature can make all-things-considered judgments and act intelligently in light of all the facts at its disposal. In general, it would be inefficient to have a sense or cognitive capacity that was not integrated with the others, because the information from this sense or cognitive capacity would be unable to figure in predictions and plans based on an assessment of all the available information.

5. *A BISA's mental processing can be understood via functional decomposition.* As complex as alien superintelligence may be, humans may be able to use the method of functional decomposition as an approach to understanding it. We've seen that a key feature of computational approaches to the brain is that cognitive and perceptual capacities are understood by decomposing the particular capacity into their causally organized parts, which themselves can be understood in terms of the causal organization of their parts. This is the method of functional decomposition, and it is a key explanatory method in cognitive science. It is difficult to envision a complex thinking machine not having a program consisting of causally interrelated elements, each of which consists of causally organized elements.

In short, the superintelligent AI's processing may make some sense to us, and developments from cognitive science may yield a glimmer of understanding into the complex mental lives of certain BISAs. All this being said, superintelligent beings are by definition beings that are superior to humans in every domain. Although a creature can have superior processing that still basically makes sense to us, it may be that a given superintelligence is so advanced that we cannot understand any of its computations whatsoever. It may be that any truly advanced civilization will have technologies that will be indistinguishable from magic, as Arthur C. Clarke suggested.[20]

In this chapter, we've zoomed away from Earth, situating mind-design issues in a cosmic context. I've illustrated that the issues we Earthlings are facing today may not be unique to Earth. In fact, discussions of superintelligence on Earth, together with research in cognitive science, helped inform speculations about what superintelligent alien minds might be like. We've also seen that our earlier discussion of synthetic consciousness is relevant as well.

It is also worth noting that as members of these civilizations develop the technology to enhance their own minds, these cultures may confront the same perplexing issues of personal identity that we discussed earlier. Perhaps the most technologically advanced civilizations are the most metaphysically daring ones, as Mandik had suggested. These are the superintelligences that didn't stall their own enhancements based on concerns about survival. Or perhaps they were concerned about personal identity and found a clever—or not so clever—way around it.

In what follows, we will descend back to Earth, delving into issues that relate to patternism. It is now time to explore a dominant view of the mind that underlies transhumanism and fusion-optimism. Many transhumanists, philosophers of mind, and cognitive scientists have appealed to a conception of the mind in which the mind is software. This is often expressed by the slogan: "the mind is the software the brain runs." It is now time to ask: Is this view of nature of the mind well founded? If our universe is stocked full of alien superintelligences, it is all the more important to consider whether the mind is software.

CHAPTER EIGHT

IS YOUR MIND A SOFTWARE PROGRAM?

I think the brain is like a programme . . . so it's theoretically possible to copy the brain onto a computer and so provide a form of life after death.

STEPHEN HAWKING[1]

One morning I awoke to a call from a *New York Times* reporter. She wanted to talk about Kim Suozzi, a 23-year-old who had died of brain cancer. A cognitive-science major, Kim was eagerly planning for graduate school in neuroscience. But the day she learned she had an exciting new internship, she also learned she had a brain tumor. She posted on Facebook: "Good news: got into The Center for Behavioral Neurosciences' BRAIN summer program. . . . Bad news: a tumor got into my BRAIN."[2]

In college, Kim and her boyfriend, Josh, had shared a common passion for transhumanism. When conventional treatments failed, they turned to cryonics, a medical technique that uses ultracold temperatures to preserve the brain upon death. Kim and Josh hoped to make the specter of death a temporary visitor. They were banking on the possibility that her brain could be revived at some point in the distant future, when there was a cure for her cancer and a means to revive cryonically frozen brains.

Kim Suozzi during an interview at Alcor, next to the containers where she and others are now frozen. (Alcor)

So Kim contacted Alcor, a nonprofit cryopreservation center in Scottsdale, Arizona. She launched a successful online campaign to get the eighty thousand dollars needed for the cryopreservation of her head. To facilitate the best possible cryopreservation, Kim was advised to spend the last weeks of her life near Alcor. So Kim and Josh moved to a hospice in Scottsdale. In her last weeks, she denied herself food and water to hasten her death, so the tumor would not further ravage her brain.[3]

Cryonics is controversial. Cryopreservation is employed in medicine to maintain human embryos and animal cells for as long as three decades.[4] But when it comes to the brain, cryopreservation is still in its infancy, and it is unknown whether someone cryopreserved using today's incipient technology could ever be revived. But Kim and Josh had weighed the pros and cons carefully.

Sadly, although Kim would never know this, her cryopreservation did not go smoothly. When the medical scans of her brain arrived, they revealed that the cryoprotectant only reached the outer portion of her brain, possibly due to vascular impairment from ischemia, leaving the remainder vulnerable to ice damage.[5] Given the damage, the author of the *New York Times* article, Amy Harmon, considered the suggestion that once uploading technology becomes available, Kim's brain be uploaded into a computer program. As she noted, certain cryopreservation efforts are turning to uploading as a means of digitally preserving the brain's neural circuits.[6]

Harmon's point was that uploading technology might benefit Kim and, more generally, those patients whose cryopreservation and illness may have damaged too much of the brain for a biological revival. The idea was that, in Kim's case, the damaged parts of the biological brain could be repaired digitally. That is, the program that her brain was uploaded to could include algorithms carrying out computations that the missing parts were supposed to achieve. And this computer program—this was supposed to be Kim.[7]

Oh boy, I thought. As a mother of a daughter only a few years younger than Kim, I had trouble sleeping that night. I kept dreaming of Kim. It was bad enough that cancer stole her life. It is one thing to cryopreserve and revive someone; there are scientific obstacles here, and Kim knew the risks. But uploading is another issue entirely. Why see uploading as a means of "revival"?

Kim's case makes all our abstract talk of radical brain enhancements so much more real. Transhumanism, fusion-optimism, artificial consciousness, postbiological extraterrestrials—it all sounds so science fiction–like. But Kim's example illustrates that, even here on Earth, these ideas are altering lives. Stephen Hawking's remarks voice an understanding of the mind that is in the air nowadays: the view that mind is a program. The *New York Times* piece reported that Kim herself had this view of the mind, in fact.[8]

Chapters Five and Six urged that uploading is far-fetched, however. It seems to lack clear support from theories of personal identity. Even modified patternism didn't support uploading. To survive uploading, your mind would have to transfer to a new location, outside your brain, through an unusual process in which information about every molecule in your brain is sent to a computer and converted into a software program. Objects that we ordinarily encounter do not "jump" across spacetime to new locations in this way. No single molecule in your brain moves to the computer, but somehow, as if by magic, your mind is supposed to transfer there.[9]

This is perplexing. For the transfer to happen, the mind must be radically unlike ordinary physical objects. My coffee cup is here, next to my laptop; when it moves, it follows a path through spacetime. It isn't dismantled, measured, and then, across the globe somewhere, configured with new components that mirror its measurements. And if it were, we wouldn't think it was the same cup, but a replica.

Furthermore, recall the reduplication problem (see Chapter Six). For instance, suppose you try to upload, and consider a scenario in which your brain and body survive the scan, as may be the case with more sophisticated uploading procedures. Suppose your upload is downloaded into an android body that looks just like you, seeming human. Feeling curious, you decide to meet your upload in a bar. As you sip a glass of wine with your android double, the two of you debate who is truly the original—who is truly *you*. The android argues convincingly that it is the real you, for it has all your memories and even remembers the beginning of the surgical procedure in which you were scanned. Your doppelgänger even asserts that it is conscious. This may be true, for we saw that if the upload is extremely precise, it may very well have a conscious mental life. But that doesn't mean it is you, for you are sitting right across from it in the bar.

In addition, if you really uploaded, you would be in principle downloadable to multiple locations at once. Suppose a hundred copies of you were downloaded. You would be *multiply located,* that is, you would be in multiple places at the same time. This is an unusual view of the self. Physical objects can be located in different places at different times, but not at the same time. We seem to be objects, albeit of a special kind: We are living, conscious beings. For us to be an exception to the generality about the behavior of macroscopic objects would be stupendous metaphysical luck.[10]

THE MIND AS THE SOFTWARE OF THE BRAIN

Such considerations motivate me to resist the siren song of digital immortality, despite my broadly transhumanist views. But what if Hawking and the others are right? What if we are lucky, because the mind truly is a kind of software program?

Suppose that Will Castor, the scientist who develops uploading in the movie *Transcendence* and becomes the first test case, was presented with the doubts raised in the last section. We tell him that the copy is not the same as the original. It is unlikely a mere information stream, running on various computers, would truly be him. He might offer the following reply:

> *The Software Response.* Uploading the mind is like uploading software. Software can be uploaded and downloaded across great distances in seconds, and can even be downloaded to multiple locations at once. We are not like ordinary physical objects at all—our minds are instead programs. So if your brain is scanned under ideal conditions, the scanning process copies your neural configuration (your "program" or

"informational pattern"). You can survive uploading insofar as your pattern survives.

The software response draws from a currently influential view of the nature of the mind in cognitive science and philosophy of mind that regards the mind as being a software program—a program that the brain runs.[11] Let's call this position "the Software View." Many fusion-optimists appeal to the Software View, along with their patternism. For instance, the computer scientist Keith Wiley writes, in response to my view:

> The mind is not a physical object at all and therefore properties of physical objects (continual path through space and time) need not apply. The mind is akin to what mathematicians and computer scientists call 'information,' for brevity a nonrandom pattern of data.[12]

If that is right, your mind can be uploaded and then downloaded into a series of different kinds of bodies. This is colorfully depicted in Rudy Rucker's dystopian novel *Software*, where the character runs out of money to pay for decent downloads and, out of desperation, dumps his consciousness into a truck. Indeed, perhaps an upload wouldn't even need to be downloaded at all. Perhaps it can just reside somewhere in a computer simulation, as in the classic film *The Matrix*, in which the notorious Smith villain has no body at all, residing solely in the Matrix—a massive computer simulation. Smith is a particularly powerful software program. Not only can he appear anywhere in the Matrix in pursuit of the good guys, he can be in multiple locations at once. At various points in the movie, Neo even finds himself fighting hundreds of Smiths.

As these science fiction stories illustrate, the Software View seems natural in the age of the Internet. Indeed, elaborations of it even describe the mind with expressions like "downloads" "apps" and "files." As Steven Mazie at *Big Think* puts it:

> Presumably you'd want to Dropbox your brain file (yes, you'll need to buy more storage) to avoid death by hard-drive crash. But with suitable backups, you, or an electronic version of you, could go on living forever, or at least for a very, very long time, "untethered," as Dr. Schneider puts it, "from a body that's inevitably going to die."[13]

Another proponent of patternism is the neuroscientist and head of the Brain Preservation Foundation, Ken Hayworth, who is irked by my critique of patternism. To him it is apparently really obvious that the mind is a program:

> It always boggles my mind that smart people continue to fall into this philosophical trap. If we were discussing copying the software and memory of one robot (say R2D2) and putting it into a new robot body would we be philosophically concerned about whether it was the 'same' robot? Of course not, just as we don't worry about copying our data and programs from an old laptop to a new one. If we have two laptops with the same data and software do we ask if one can 'magically' access the other's RAM? Of course not.[14]

So, is the Software View correct? No. The software approach to the mind is deeply mistaken. It is one thing to say that the brain is computational; this is a research paradigm in cognitive science that I am quite fond of (see, for instance, my earlier book, *The Language of Thought*). Although the Software View is often taken as being part and parcel of the computational approach to the brain, many metaphysical approaches to the nature of mind are compatible with a computational approach to the brain.[15] And, as I'll explain shortly, the view that the mind or self is software is one we should do without.

Before I launch into my critique, let me say a bit more about the significance of the Software View. There are at least two reasons the issue is important. First, if the Software View is correct, patternism is more plausible than Chapters Five and Six indicated. My objections involving spatiotemporal discontinuity and reduplication can be dismissed, although other problems remain, such as deciding when an alteration in a pattern is compatible with survival and when it is not.

Second, if the Software View is correct, it would be an exciting discovery, because it would provide an account of the nature of the mind. In particular, it might solve a central philosophical puzzle known as the Mind-Body Problem.

THE MIND-BODY PROBLEM

Suppose that you are sitting in a cafe studying right before a big presentation. All in one moment, you taste the espresso you sip, feel a pang of anxiety, consider an idea, and hear the scream of the espresso machine. What is the nature of these thoughts? Are they just a matter of physical states of your brain, or are they something more? Relatedly, what is the nature of your mind? Is your mind just a physical thing, or is it something above and beyond the configuration of particles in your brain?

These questions pose the Mind-Body Problem. The problem is where to situate mentality within the world that science investigates. The Mind-Body Problem is closely related to the aforementioned Hard Problem of Consciousness, the puzzle of why physical processes are accompanied by subjective feeling. But the focus of the Hard Problem is consciousness, whereas the Mind-Body Problem focuses on mental states more generally, even nonconscious mental states. And instead of asking why these states must exist, it seeks to determine how they relate to what science investigates.

Contemporary debates over the Mind-Body Problem were launched more than 50 years ago, but some classic positions began to emerge as early as the pre-Socratic Greeks. The problem is not getting any easier. There are some fascinating solutions, to be sure. But as with the debate over personal identity, there are no uncontroversial ones in sight. So, does the Software View solve this classic philosophical problem? Let's consider some influential positions on the problem and see how the Software View compares.

Panpsychism

Recall that panpsychism holds that even the smallest layers of reality have experience. Fundamental particles have minute levels of consciousness, and in a watered-down sense, they are subjects of experience. When particles are in extremely sophisticated configurations—such as when they are in nervous systems—more sophisticated, recognizable forms of consciousness arise. Panpsychism may seem outlandish, but the panpsychist would respond that their theory actually meshes with fundamental physics, because experience is the underlying nature of the properties that physics identifies.

Substance Dualism

According to this classic view, reality consists of two kinds of substances, physical things (e.g., brains, rocks, bodies) and nonphysical ones (i.e., minds, selves, or souls). Although you personally may reject the view that there's an immaterial mind or soul, science alone cannot rule it out. The most influential philosophical substance dualist, René Descartes, thought that the workings of the nonphysical mind corresponded with the workings of the brain, at least during one's lifetime.[16] Contemporary substance dualists offer sophisticated nontheistic positions, as well as intriguing and equally sophisticated theistic ones.

Physicalism (or Materialism)

We discussed physicalism briefly in Chapter Five. According to physicalism, the mind, like the rest of reality, is physical. Everything is either made up of something that physics describes or is a fundamental property, law, or substance figuring in a physical theory. (Here, by "physical theory," physicalists tend to gesture toward the content of the final theory of everything that a completed physics uncovers, whatever that is.) There are no immaterial minds or souls, and all of our thoughts are ultimately just physical phenomena. This position has been called "materialism," but it is now more commonly called "physicalism." Because there is no second immaterial realm, as substance dualism claimed, physicalism is generally regarded as a form of monism—the claim that there is one fundamental type of category to reality—in this case, the category of physical entities.

Property Dualism

The point of departure for this position is the hard problem of consciousness. Proponents of property dualism believe that the best answer to the question, "why does consciousness need to exist?" is that consciousness is a fundamental feature of certain complex systems. (Paradigmatically, such features emerge from the biological brain, but perhaps one day, synthetic intelligences will have such features as well). Property dualists, like substance dualists, claim that reality divides into two distinct realms. But property dualists reject the existence of souls and immaterial minds. Thinking systems are physical things, but they have non-physical properties (or features). These nonphysical features are basic building blocks of reality, alongside fundamental physical properties, but unlike panpsychism, these basic features are not microscopic—they are features of complex systems.

Idealism

Idealism is less popular than the other views, but it has been historically significant. Idealists hold that fundamental reality is mind-like. Some advocates of this view are panpsychists, although a panpsychist can also reject idealism, claiming that there is more to reality than just minds or experiences.[17]

There are many intriguing approaches to the nature of mind, but I've focused on the most influential. Should the reader wish to consider solutions to the Mind-Body Problem in more detail, there are several excellent introductions available.[18] Now that we've considered these positions, let us turn back to the Software View and see how it fares.

ASSESSING THE SOFTWARE VIEW

The Software View has two initial flaws, both of which can be remedied, I believe. First, not all programs are the sorts of things that have minds. The Amazon or Facebook app on your smart-phone doesn't have a mind, at least if we think of minds in the normal sense (i.e., as something only highly complex systems, such as brains, have). If minds are programs, they are programs of a very special sort, having layers of complexity that fields like psychology and neuroscience find challenging to describe. A second issue is that, as we've seen, consciousness is at the heart of our mental lives. A zombie program—a program incapable of having experience—just isn't the sort of thing that has a mind.

Yet these points are not decisive objections, for if proponents of the Software View agree with one or both of these criticisms, they can qualify their view. For instance, if they agree with both criticisms, they can restrict the Software View in the following way:

> Minds are programs of a highly sophisticated sort, which are capable of having conscious experiences.

But adding some fine print doesn't fix the deeper problems I will raise.

To determine whether the Software View is plausible, let us ask: What is a program? As the image on the following page indicates, a program is a list of instructions in lines of computer code. The lines of code are instructions in a programming language that tell the computer what tasks to do. Most computers can execute several programs and, in this way, new capacities can be added or deleted from the computer.

A line of code is like a mathematical equation. It is highly abstract, standing in stark contrast with the concrete physical world around you. You can throw a rock. You can lift a coffee cup. But just try to throw an equation. Equations are abstract entities; they are not situated in space or time.

```
95          <div class="container">
            <h1>One more for good measure.
96          <p>Cras justo odio, dapibus ac facilisis in,
97          .</p>
98          <p><a class="btn btn-lg btn-primary" href=
99          </div>
100       </div>
101     </div>
102     <a class="left carousel-control" href="#myCarousel" role=
103       <span class="glyphicon glyphicon-chevron-left" aria-hidden=
104       <span class="sr-only">Previous</span>
105     </a>
106     <a class="right carousel-control" href="#myCarousel" role=
107       <span class="glyphicon glyphicon-chevron-right" aria-hidden=
108       <span class="sr-only">Next</span>
109     </a>
110   </div><!-- /.carousel -->
111
112   <!--Featured Content Section-->
113
114   <div class="container">
115     <div class="row">
116       <div class="col-md-4"></div>
          <div class="col-md-4"> <h2> FEATURED CONTENT </h2>
          <div class="col-md-4"></div>
          <div class="
```

Now that we appreciate that a program is abstract, we can locate a serious flaw in the Software View. If your mind is a program, then it is just a long sequence of instructions in a programming code. The Software View is saying the mind is an abstract entity. But think about what this means. The field of philosophy of mathematics studies the nature of abstract entities like equations, sets, and programs. Abstract entities are said to be *nonconcrete: They are nonspatial, nontemporal, nonphysical, and acausal.* The inscription "5" is here on this page, but the actual number, as opposed to the inscription, isn't located anywhere. Abstract entities are not located in space or time, they are not physical objects, and they do not cause events to occur in the spatiotemporal manifold.

How can the mind be an abstract entity, like an equation or the number 2? This seems to be a category mistake. We are spatial beings and causal agents; our minds have states that cause us to act in the concrete world. And moments pass for us—we are temporal beings. So, your mind is not an abstract entity like a program. Here, you may suspect that programs are able to act in the world. What about the last time your computer crashed, for instance? Didn't the program cause the crash? But this confuses the program with one of its instantiations. At one instance, say, when Windows is running, the Windows program is implemented by physical states within a particular machine. The machine and its related process are what crashes. We might speak of a program crashing, but on reflection, the algorithm or lines of code (i.e., the program) do not literally crash or cause a crash. The electronic states of a particular machine cause the crash.

So the mind is not a program. And there is still reason to doubt that uploading the mind is a true means for Kim Suozzi, or others, to survive. As I've been stressing throughout the second part of this book, assuming that there are selves that persist over time, biologically based enhancements that gradually restore and cautiously enhance the functioning of the biological brain are a safer route to longevity and enhanced mental abilities, even if the technology to upload a complete human brain is developed. Fusion-optimists tend to endorse both rapid alterations in psychological continuity or radical changes in one's substrate. Both types of enhancements seem risky, at least if one believes that there is such a thing as a persisting self.

I emphasized this in Chapters Five and Six as well, although there, my rationale did not concern the abstract nature of the Software View. There, my caution stemmed from the controversy in metaphysics over which, if any, competing theories of the nature of the person are correct. This left us adrift concerning whether radical, or even moderate, enhancements are compatible with survival. We can now see that just as patternism about the survival of the person is flawed, so too, the related Software View is problematic. The former runs afoul of our understanding the nature of personhood, while the latter ascribes a physical significance to abstractions that they do not possess.

I'd like to caution against drawing a certain conclusion from my rejection of the Software View, however. As I've indicated, the computational approach to the mind in cognitive science is an excellent explanatory framework.[19] But it does not entail the view that the mind is a program. Consider Ned Block's canonical paper, "The Mind as the Software of the Brain."[20] Aside from its title, which I obviously disagree with, it astutely details many key facets of the view that the brain is computational. Cognitive capacities, such as intelligence and working memory, are explainable via the method of functional decomposition; mental states are multiply realizable; and the brain is a syntactic engine driving a semantic one. Block is accurately describing an explanatory framework in cognitive science by isolating key features of the computational approach to the mind. None of this entails the metaphysical position that the mind is a program, however.

The Software View isn't a viable position, then. But you might wonder whether the transhumanist or fusion-optimist could provide a more feasible computationalist approach to the nature of the mind. As it happens, I do have a further suggestion. I think we can formulate a transhumanist-inspired view in which minds are not programs per se, but program instantiations—a given run of a program. We will then need to ask whether this modified view is any better than the standard Software View.

COULD LIEUTENANT COMMANDER DATA BE IMMORTAL?

Consider Lieutenant Commander Data, the android from *Star Trek: The Next Generation*. Suppose he finds himself in an unlucky predicament, on a hostile planet, surrounded by aliens that are about to dismantle him. In a last-ditch act of desperation, he quickly uploads his artificial brain onto a computer on the *Enterprise*. Does he survive? And could he, in principle, do this every time he's in a jam, so that he'd be immortal?

If I'm correct that neither Data's nor anyone else's mind is a software program, this has bearing on the question of whether AIs, including uploads, could achieve immortality or, rather, whether they could achieve what we might call "functional immortality." (I write "functional immortality," because the universe may eventually undergo a heat death that no life can escape. But I'll ignore this technicality in what follows.)

It is common to believe that an AI could achieve functional immortality by creating backup copies of itself and thus transfer its consciousness from one computer to the next when an accident happens. This view is encouraged by science fiction stories, but I suspect it is mistaken. Just as it is questionable whether a human could achieve functional immortality by uploading and downloading herself, so, too, we can question whether an AI would genuinely survive. Insofar as a particular mind is not a program or abstraction but a concrete entity, a particular AI mind is vulnerable to destruction by accident or the slow decay of its parts, just as we are.

This is hardly an obvious point. It helps to notice that there is an ambiguity as to whether "AI" refers to a particular AI (an individual being) or to a type of AI system (which is an abstract entity). By analogy, "the Chevy Impala" could mean the beat-up car you bought after college, or it could mean the type of car (i.e., the make and model). That would endure even after you scrapped your car and sold it for parts. So, it is important to disambiguate claims about survival. Perhaps, if one wants to speak of types of programs as "types of mind," the types could be said to "survive" uploading, according to two watered-down notions of survival. First, at least in principle, a machine that contains a high-fidelity copy of an uploaded human brain can run the same program as that brain did before it was destroyed by the uploading procedure. The type of mind "survives," although no single conscious being persists. Second, a program, as an abstract entity, is timeless. It does not cease to exist, because it is not a temporal being. But this is not "survival" in a serious sense. Particular selves or minds do not survive in either of these two senses.

This is all highly abstract. Let's return to the example of Lieutenant Commander Data. Data is a particular AI, and as such, he is vulnerable to destruction. There may be other androids of this type (individual AIs themselves), but their survival does not ensure the survival of Data, it just ensures the "survival" of Data's type of mind. (I write "survival" in scare quotes to indicate that I am referring to the aforementioned watered-down sense of survival.)

So there Data is, on a hostile planet, surrounded by aliens that are about to destroy him. He quickly uploads his artificial brain onto a computer on the *Enterprise*. Does he survive or not? On my view, we now have a distinct instance (or, as philosophers say, a "token") of the type of mind *Data* being run by that particular computer. We could ask: Can that token survive the destruction of the computer by uploading again (i.e., transferring the mind of that token to a different computer)? No. Again, uploading would merely create a different token of the same type. An individual's survival depends on where things stand at the token level, not at the level of types.

It is also worth underscoring that a particular AI could still live a very long time, insofar as its parts are extremely durable. Perhaps Data could achieve functional immortality by avoiding accidents and having his parts replaced as they wear out. My view is compatible with this scenario, because Data's survival in this case does not happen by transferring his program from one physical object to another. On the assumption that one is willing to grant that humans survive the gradual replacement of their parts over time, why not also grant it in the case of AIs? Of course, in Chapter Five, I emphasized that it is controversial whether persons survive the replacement of parts of their brains; perhaps the self is an illusion, as Derek Parfit, Friedrich Nietzsche, and the Buddha have suggested.

IS YOUR MIND A PROGRAM INSTANTIATION?

The central claim of my discussion of Data is that survival is at the token level. But how far can we push this observation? We've seen that the mind is not a program, but could it be the *instantiation* of a program—the thing that runs the program or stores its informational pattern? Something that instantiates a program is a concrete entity—paradigmatically, a computer, although technically, a program instantiation involves not just the computer's circuitry but also the physical events that occur in the computer when the program is running. The pattern of matter and energy in the system corresponds, in possibly nontrivial ways, to elements of the program (e.g., variables, constants).[21] Let us call this position *the Software Instantiation View of the Mind*:

The Software Instantiation View of the Mind (SIM)

The mind is the entity running the program (where a program is the algorithm that the brain implements, something in principle discoverable by cognitive science).

This new position does not serve the fusion-optimist well, however. This is not a view that is accurately expressed by the slogan: "The mind is the software of the brain." Rather, this view is claiming that mind is the entity running the program. To see how different SIM is from the Software View, notice that SIM doesn't suggest that Kim Suozzi can survive uploading; my earlier concerns involving spatiotemporal discontinuities still apply. As with modified patternism, each upload or download is not the same person as the original, although it has the same program.

The above definition specifies that the program runs on a brain, but we can easily broaden it to other substrates, such as silicon-based computers:

The Software Instantiation Approach to the Mind (SIM*)

The mind is the entity running the program (where a program is the algorithm that the brain or other cognitive system implements, something in principle discoverable by cognitive science).

SIM*, unlike the original Software View, avoids the category mistake of viewing the mind as abstract. But like the original view and the related patternist position, it draws from the computational approach to the brain in cognitive science.

Does SIM* provide a substantive approach to the Mind-Body Problem? Consider that it hasn't told us about the underlying metaphysical nature of the thing that runs the program (i.e., the mind). So it is uninformative. For the Software Instantiation approach to serve as an informative theory of the nature of the mind, it needs to take a stand on each of the aforementioned positions on the nature of the mind.

Consider *panpsychism,* for instance. Does the system that instantiates the program consist of fundamental elements that have their own experiences? SIM* leaves this question open. Furthermore, SIM* is also compatible with physicalism, the view that everything either is made of something that physics describes or is a fundamental property, law, or substance figuring in a physical theory.

Property dualism is also compatible with the mind being a program instantiation. For instance, consider the most popular version of the view, David Chalmers's naturalistic property dualism. According to Chalmers, features like *seeing the rich hues of a sunset,* or *smelling the aroma of the espresso* are properties that emerge from complex structures. Unlike panpsychism, these fundamental consciousness properties are not found in fundamental particles (or strings)—they are at a higher level, inhering in highly complex systems. Nevertheless, these properties are basic features of reality.[22] So physics will be incomplete, no matter how sophisticated it gets, because in addition to physical properties, there are novel fundamental properties that are nonphysical. Notice that SIM* is compatible with this view, because the system that runs the program could have certain properties that are nonphysical and are basic features of reality in their own right.

Recall that substance dualism claims that reality consists of two kinds of substances: physical entities (e.g., brains, rocks, bodies) and nonphysical ones (i.e., minds, selves, or souls). Nonphysical substances could be the sort of entities that run a program, so SIM* is compatible with substance dualism. This may sound strange, so it pays to consider how such a story would go; the details depend on the kind of substance dualism that is in play.

Suppose a substance dualist says, as Descartes did, that the mind is wholly outside of spacetime. According to Descartes, although the mind isn't in spacetime, throughout a person's life, one's mind is still capable of causing states in one's brain, and vice versa.[23] (How does it do this? I'm afraid Descartes never gave a viable account, claiming, implausibly, that mind-brain interactions happened in the pineal gland.)

How would a program implementation view be compatible with Cartesian dualism? In this case, the mind, if a program instantiation, would be a nonphysical entity that is outside of spacetime. During a person's worldly life, the mind causes states of the brain. (Notice that the nonphysical mind is not an abstract entity, however, as it has causal and temporal properties. Being nonspatial is a necessary condition of being abstract, but it is not a sufficient condition.) We might call this view *Computational Cartesianism*. This may sound odd, but experts on functionalism, like the philosopher Hilary Putnam, have long recognized that computations of a Turing machine can be implemented in a Cartesian soul.[24]

The picture that Computational Cartesianism offers of mind-body causation is perplexing, but so was the original Cartesian view that the mind, although nonspatiotemporal, somehow stands in a causal relationship with the physical world.

Not all substance dualisms are this radical, in any case. For instance, consider non-Cartesian substance dualism, a view held by E. J. Lowe. Lowe held that the self is distinct from the body. But in contrast to Cartesian dualism, Lowe's dualism doesn't claim either that the mind is separable from the body or that it is nonspatial. It allows that the mind may not be able to exist without a body and that, being spatiotemporal, it possesses properties, such as shape and location.[25]

Why did Lowe hold this position? Lowe believed that the self is capable of survival across different kinds of physical substrates, so it has different persistence conditions than the body. We've seen that such claims about persistence are controversial. But you do not need to share Lowe's intuitions about persistence; the point here is simply to raise a different, non-Cartesian, substance dualist position. SIM* is compatible with non-Cartesian substance dualism, because a program instantiation could be this sort of nonphysical mind as well. This position is harder to dismiss than Cartesianism, as minds are part of the natural world. Yet again, the Software Instantiation View remains silent.

In essence, although SIM* does not venture the implausible claim that the mind is abstract, it tells us little about the nature of mind, except that it is something that runs a program. Anything could do that, in principle—Cartesian minds, systems made of fundamental experiential properties, and so on. This isn't much of a position on the Mind-Body Problem then.

At this point, perhaps a proponent of SIM* would say that they intend to be making a different sort of metaphysically substantive claim, however, one that concerns the persistence of the mind over time. Perhaps they hold the following view:

Being a program instantiation of type T is an *essential property* of one's mind, without which the mind couldn't persist.

Recall that your contingent properties are ones that you can cease to have and still continue to exist. For instance, you may change your hair color. Your essential properties, in contrast, are those that are essential to you. Earlier, we considered the debate over the persistence of persons; in a similar vein, the proponent of SIM* can say that being an instantiation of program T is essential to one's continuing to have the mind one has, and one's mind would cease to exist if T changed to a different program, P.

Is this a plausible position? One problem is that a program is just an algorithm, so if any lines of the algorithm change, the program changes. The brain's synaptic connections are constantly altered to reflect new learning, and when you learn something, such as a new skill, this leads to changes in your "program." But if the program changes, the mind ceases to exist, and a new one begins. Ordinary learning shouldn't result in the death of your mind.

Proponents of the program instantiation view can respond to this objection, however. They could say that a program can exhibit historical development, being expressed by an algorithm of type T_1 at time t, and, at a later time, be expressed by an algorithm that is a modified version of T_1, T_2. Although, technically, T_1 and T_2 are distinct, consisting of at least some different instructions, T_1 is an ancestor of T_2. So the program continues. On this view, the person is the instantiation of a certain program, and the program can change in certain ways but still remain the same program.

RIVERS, STREAMS, AND SELVES

Notice that this is just the aforementioned transhumanist "patternist" view, modified to hold that the person is not the pattern but the *instantiation* of the pattern. Earlier, we discussed a similar view, called "modified patternism." So we've come full circle. Recall Kurzweil's remark:

> I am rather like the pattern that water makes in a stream that rushes past the rocks in its path. The actual molecules of water change every millisecond, but the pattern persists for hours or even years.[26]

Of course, Kurzweil knows that over time, the pattern will change. After all, this is a passage from his book on becoming posthuman during the singularity. Kurzweil's remark may strike a chord with you: In an important sense, you seem to be the same person you were a year ago, despite the changes to your brain, and perhaps you even could survive the loss of many of your memories or the addition of some new neural circuitry, such as an enhancement to your working memory system. Perhaps, then, you are like the river or stream.

The irony is that the metaphor of a river was used by the pre-Socratic philosopher Heraclitus to express the view that reality is flux. Persisting things are an illusion, including the permanence of a persisting self or mind. Thousands of years ago, Heraclitus wrote: "No man ever steps in the same river twice, for it's not the same river and he's not the same man."[27]

Yet Kurzweil is saying that the self survives the flux of change. The challenge for modified patternists is to resist Heraclitus's move: to show that there is a permanent self, against the backdrop of continual change, rather than the mere illusion of permanence. Can the modified patternists impose the permanence of the self on the Heraclitan flux of ever-changing molecules in the body?

Here, we run up against a familiar problem. Without a firm handle on when a pattern implementation does or does not continue, we do not have good reason to appeal to the Software Instantiation View either. In Chapter Five, we asked: If you are the instantiation of a certain pattern, what if your pattern shifts? Will you die? The extreme cases, like uploading, seemed clear. However, mere everyday cellular maintenance by nanobots to overcome the slow effects of aging would perhaps not affect the identity of the person. But we saw that the middle-range cases are unclear. Remember, the path to superintelligence may very well be a path through middle-range enhancements that add up, over time, to major changes to one's cognitive and perceptual makeup. Further, as we observed in Chapter Five, selecting a boundary seems arbitrary, for once a boundary is selected, an example can be provided that suggests the boundary should be pushed outward.

So then, if proponents of the Software Instantiation View have a position about persistence in mind, the same old issue rears its ugly head. We have indeed come full circle, and we are left with an appreciation of how perplexing and controversial these mysteries about the nature of the mind and self are. And this, dear reader, is where I want to leave you. For the future of the mind requires appreciating the metaphysical depth of these problems.

Now let's bring our discussion home, summing things up by returning to the Suozzi case.

RETURNING TO ALCOR

Three years after Kim's death, Josh gathered her special belongings and returned to Alcor. He was making good on a promise to her, delivering her things where she can find them if she is brought back to life.[28] Frankly, I wish the conclusion of this chapter had been different. If the Software View was correct, then at least in principle, minds could be the sort of thing that can be uploaded, downloaded, and rebooted. This would allow an afterlife for the brain, if you will—a way in which a mind, like Kim's, could survive the death of the brain. Yet our reflections revealed that the Software View turned minds into abstract objects. So we considered a related view, one in which the mind is a program instantiation. We then saw that the Program Instantiation View does not support uploading either, and although it is an interesting approach, it is too uninformative, from a metaphysical standpoint, to be much of an approach to the nature of mind.

Although I do not have access to the medical details of Kim's cryopreservation, I find hope in the *New York Times* report that there was imaging evidence that the outer layers of Kim's brain were successfully cryopreserved. As Harmon notes, the brain's neocortex seems central to who we are, being key to memory and language.[29] So perhaps a biologically-based reconstruction of the damaged parts would be compatible with the survival of the person. For instance, I've observed that, even today, there is active work on hippocampal prosthetics; perhaps parts of the brain like the hippocampus are rather generic, and replacing them with biological or even AI-based prosthetics doesn't change who one is.

Of course, I've stressed throughout this book that there is tremendous uncertainty here, due to the perplexing and controversial nature of the personal identity debate. But Kim's predicament is not like that of someone seeking an optional brain enhancement, such as a shopper strolling into our hypothetical Mind Design Center. A shopper browsing a menu of enhancements can comfortably reject an enhancement because it strikes them as too risky, but a patient on the brink of death, or who requires a prosthetic to be cryogenically revived, may have little to lose, and everything to gain, in pursuing a high-risk cure.

Desperate times call for desperate measures. A decision to use one, or even several, neural prosthetics, to facilitate Kim's revival seems rational, if the technology is perfected. In contrast, I have no confidence that resorting to uploading her brain would be a form of revival. Uploading, at least as a means of survival, rests on flawed conceptual foundations.

Should uploading projects be scrapped then? Even if uploading technology doesn't fulfill its original promise of digital immortality, perhaps it can nevertheless benefit our species. For instance, a global catastrophe may make Earth inhospitable to biological life forms, and uploading may be a way to preserve the human way of life and thinking, if not the actual humans themselves. And if these uploads are indeed conscious, this could be something that members of our species come to value, when confronted with their own extinction. Furthermore, even if uploads aren't conscious, the use of simulated human minds for space travel could be a safer, more efficient way of sending intelligent beings into space than sending a biological human. The public tends to find manned missions to space exciting, even when robotic missions seem more efficient. Perhaps the use of uploaded minds would excite the public. Perhaps these uploads could even run terraforming operations on inhospitable worlds, readying the terrain for biological humans. You never know.

In addition, brain uploading could facilitate the development of brain therapies and enhancements that could benefit humans or nonhuman animals, because uploading part or all of a brain could help generate a working emulation of a biological brain that we could learn from. AI researchers who aim to build AIs that rival human-level intelligence may find it a useful means of AI development. Who knows, perhaps AI that is descended from us will have a greater chance of being benevolent toward us.

Finally, some humans will understandably want digital doubles of themselves. If you found out that you were going to die soon, you may wish to leave a copy of yourself to communicate with your children or complete projects that you care about. Indeed, the personal assistants—the Siris and Alexas of the future—might be uploaded copies of deceased humans we have loved deeply. Perhaps our friends will be copies of ourselves, tweaked in ways we find insightful. And perhaps we will find that these digital copies are themselves sentient beings, deserving to be treated with dignity and respect.

CONCLUSION: THE AFTERLIFE OF THE BRAIN

At the heart of this book is a dialogue between philosophy and science. We've seen that the science of emerging technologies can challenge and expand our philosophical understanding of the mind, self, and person. Conversely, philosophy sharpens our sense of what these emerging technologies can achieve: whether there can be conscious robots, whether you could replace much of your brain with microchips and be confident that it would still be you, and so on.

This book has attempted, very provisionally, to explore mind design space. Although we do not know whether there will be places like Immortex or the Center for Mind Design, I wouldn't be surprised. The events of today speak for themselves: This is a time when AI is projected to replace most blue- and white-collar jobs over the next several decades, and in which there are active efforts to merge humans with machines.

I've suggested that it is not a foregone conclusion that sophisticated AIs will be conscious. Instead, I've argued for a middle-of-the-road approach, one that rejects the Chinese Room thought experiment, yet doesn't assume, based on the computational nature of the brain or the conceptual possibility of isomorphs, that sophisticated AI will be conscious. Conscious AI may not, in practice, be built, or it may not be compatible with the laws of physics to create consciousness in a different, nonbiological, substrate. But by carefully gauging whether the AIs we develop are conscious, we can approach the issue with care. And by publicly debating all this in a way that moves beyond both technophobia and the tendency to view the AIs that look superficially humanlike as conscious, we will be better able to judge whether and how conscious AI should be built. These should be choices societies make carefully, and all stakeholders must be involved.

I've further emphasized that from an ethical standpoint, it is best to assume that a sophisticated AI may be conscious, at least until we develop tests for consciousness that we can have confidence in. Any mistake could wrongly influence the debate over whether AIs might be worthy of special ethical consideration as sentient beings, and it is better to err on the side of safety. Not only could failure to recognize a machine as sentient cause needless pain and suffering, but, as films like *Ex Machina* and *I, Robot* illustrate, any failure to be charitable to AI may come back to haunt us, as they may treat us as we treated them.

Some younger readers may one day be faced with an opportunity to make mind-design decisions. If you are one of these readers, my message to you is this: Before you enhance, you must reflect on who you are. If you are uncertain as to the ultimate nature of the person, as I am, take a safe, cautious route: As much as possible, stick to gradual, biologically based therapies and enhancements, ones that mirror the sort of changes that normal brains undergo in the process of learning and maturation. Bearing in mind all the thought experiments that question more radical approaches to enhancement, and the general lack of agreement in the personal-identity debate, this cautious approach is most prudent. It is best to avoid radical, quick changes, even ones that do not alter the type of substrate on has (e.g., carbon versus silicon). It is also prudent to avoid attempts to "migrate" the mind to another substrate.

Until we know more about synthetic consciousness, we cannot be confident that transferring key mental functions to AI components is safe in parts of the brain underlying consciousness. Of course, we have yet to determine whether AI is conscious, so we do not know whether you, or perhaps more precisely, the AI copy of you, would even be a conscious being, if you tried to merge with AI.

By now, you can see that any trip to the Center for Mind Design could be vexing, even perilous. I wish I could provide you with a clear, uncontroversial path to guide you through the mind-design decisions. Instead, my message has been: As we consider mind-design decisions, we must do so, first and foremost, with a mindset of metaphysical humility. Remember the risks. The future of the mind, whether the minds are human minds or robot minds, is a matter that requires public dialogue and philosophical contemplation.

TRANSHUMANISM

Transhumanism is not a monolithic ideology, but it does have an official declaration and an organization. The World Transhumanist Association is an international nonprofit organization that was founded in 1998 by philosophers David Pearce and Nick Bostrom. The main tenets of transhumanism are stated in the Transhumanist Declaration, which is reprinted below.[1]

The Transhumanist Declaration

1. Humanity will be radically changed by technology in the future. We foresee the feasibility of redesigning the human condition, including such parameters as the inevitability of aging, limitations on human and artificial intellects, unchosen psychology, suffering, and our confinement to the planet earth.
2. Systematic research should be put into understanding these coming developments and their long-term consequences.
3. Transhumanists think that by being generally open and embracing of new technology we have a better chance of turning it to our advantage than if we try to ban or prohibit it.
4. Transhumanists advocate the moral right for those who so wish to use technology to extend their mental and physical (including reproductive) capacities and to improve their control over their own lives. We

seek personal growth beyond our current biological limitations.

5. In planning for the future, it is mandatory to take into account the prospect of dramatic progress in technological capabilities. It would be tragic if the potential benefits failed to materialize because of technophobia and unnecessary prohibitions. On the other hand, it would also be tragic if intelligent life went extinct because of some disaster or war involving advanced technologies.

6. We need to create forums where people can rationally debate what needs to be done, and a social order where responsible decisions can be implemented.

7. Transhumanism advocates the well-being of all sentience (whether in artificial intellects, humans, posthumans, or non-human animals) and encompasses many principles of modern humanism. Transhumanism does not support any particular party, politician or political platform.

This document was followed by the much longer and extremely informative Transhumanist Frequently Asked Questions, which is widely available online.[2]

ACKNOWLEDGMENTS

This book was a delight to write, and I am very grateful to those who provided feedback on the project and to the institutions that have sponsored the research for this book. Chapters Two through Four were written in the midst of a stimulating project on AI consciousness at Stanford Research Institute (SRI). Chapter Seven grew out of my research project with NASA and a fruitful series of collaborations with those at the Center for Theological Inquiry (CTI) in Princeton, NJ. Special thanks to Robin Lovin, Josh Mauldin, and Will Storrar for hosting me there.

I also owe thanks to Piet Hut at the Institute for Advanced Study (IAS) in Princeton for hosting me as a visiting member at the Institute. I've learned a good deal from the members of our weekly AI lunch group, organized by Hut and Olaf Witkowski. Edwin Turner has been a frequent collaborator, both at IAS and CTI, and I've enjoyed our joint work immensely. I also benefited from discussing these issues with the members of my AI, Mind and Society (AIMS) Group. In particular, Mary Gregg, Jenelle Salisbury, and Cody Turner deserve special thanks for their insightful remarks on chapters of this book.

Some of these chapters draw from shorter, earlier pieces that appeared in *The New York Times*, *Nautilus*, and *Scientific American*. The themes of Chapter Four were inspired and expanded from a piece in *Ethics of Artificial Intelligence* (Liao, 2020), and Chapter Six expands on material in my "Mindscan: Transcending and Enhancing the Human Brain," in *Science Fiction and Philosophy* (Schneider, 2009b). Chapter Seven draws from essays appearing in astrobiology volumes by Dick (2013) and Losch (2017).

While putting the final touches on the book, I've served as the Distinguished Scholar chair at the Library of Congress, and I am grateful to those at the Kluge Center for hosting me there, especially John Haskell, Travis Hensley, and Dan Turello. I'm also appreciative of the feedback from colleagues at the University of Connecticut, where I presented this material at our department brown bags and in a cognitive science colloquium. I'm grateful to audiences and hosts at talks at Cambridge University, the University of Colorado, Yale University, Harvard University, the University of Massachusetts, Stanford University, the University of Arizona, Boston University, Duke University, 24Hours, and the cognitive science and plasma physics departments and the Woodrow Wilson School at Princeton University.

I greatly appreciate the efforts of those who held and spoke at conferences on themes in my work. "Minds, Selves and Technology" in Lisbon, Portugal, was organized by Rob Clowes, Klaus Gardner, and Ines Hipolito. I am also grateful to the Czech Academy of Sciences for hosting the Ernst Mach Workshop in Prague in the June 2019 celebration of this book. And I thank PBS for televising my lecture on the material that became Chapter Six, and Greg Gutfeld at Fox TV for hosting me for an entire show covering the material in the book.

Stephen Cave, Joe Corabi, Michael Huemer, George Musser, Matt Rohal, and Eric Schwitzgebel sent extensive comments on the entire manuscript and were responsible for many improvements. I've also benefited tremendously from conversations on the material with John Brockman, Antonio Chella, David Chalmers, Eric Henney, Carlos Montemayor, Martin Rees, David Sahner, Michael Solomon, and Dan Turello. I am grateful to Josh Schishler for discussing his experience with Kim Suozzi with me. Special thanks to the fabulous team at

Princeton University Press (Cyd Westmoreland, Sara Henning-Stout, Rob Tempio, and all the others) and especially to Matt Rohal for editing the book with such care. (Alas, I worry that I've failed to remember someone's efforts and insights, and if this is the case, I'm sorry.)

Finally, I thank my husband, David Ronemus. Our wonderful conversations about AI helped inspire the material in the book. This book is lovingly dedicated to our children: Elena, Alex, and Ally. I would be immensely gratified if this book makes even a modest contribution toward helping younger generations negotiate the technological and philosophical challenges I've discussed.

NOTES

INTRODUCTION: YOUR VISIT TO THE CENTER FOR MIND DESIGN

1. *Contact,* film directed by Robert Zemeckis, 1997.

2. See, for example, the open letter https://futureoflife.org/ai-open-letter/, Bostrom (2014), Cellan-Jones (2014), Anthony (2017), and Kohli (2017).

3. Bostrom (2014).

4. Solon (2017).

CHAPTER ONE: THE AGE OF AI

1. Müller and Bostrom (2016).

2. Giles (2018).

3. Bess (2015).

4. Information about some of this research can be found at clinicaltrials.gov, a database of privately and publicly funded clinical studies conducted around the world. See also publicly available discussions of some of the research conducted by the Defense Advanced Research Projects Agency (DARPA), which is the emerging technologies wing of the U.S. Department of Defense: DARPA (n.d. a); DARPA (2018); *MeriTalk* (2017). See also Cohen (2013).

5. Huxley (1957, pp. 13–17). For a variety of classic papers on transhumanism, see More and Vita-More (2013).

6. Roco and Bainbridge (2002); Garreau (2005).

7. Sandberg and Bostrom (2008).

8. DARPA (n.d. b).

9. Kurzweil (1999, 2005).

CHAPTER TWO: THE PROBLEM OF AI CONSCIOUSNESS

1. Kurzweil (2005).

2. Chalmers (1996, 2002, 2008).

3. The Problem of AI Consciousness is also distinct from a classic philosophical problem called "the problem of other minds." Each of us can tell, by introspection, that we are conscious, but how can we really be sure that the other humans around

us are? This problem is a well-known form of philosophical skepticism. A common reaction to the problem of other minds is to hold that although we cannot tell with certainty that the people around us are conscious, we can infer that other normal humans are conscious, because they have nervous systems like our own, and they exhibit the same basic kinds of behaviors, such as wincing when in pain, seeking friendships, and so on. The best explanation for the behaviors of other humans is that they are also conscious beings. After all, they have nervous systems like ours. The problem of other minds is different from the Problem of AI Consciousness, however. For one thing, it is posed in the context of human minds, not machine consciousness. Furthermore, the popular solution to the problem of other minds is ineffective in the context of the Problem of AI Consciousness. For AIs do not have nervous systems like our own, and they may behave in quite alien ways. Additionally, if they do behave like humans, it may be because they are programmed to behave as if they feel, so we can't infer from their behavior that they are conscious.

4. Biological naturalism is often associated with the work of John Searle. But "biological naturalism," as used here, doesn't involve Searle's broader position about physicalism and the metaphysics of mind. For this broader position, see Searle (2016, 2017). For our purposes, biological naturalism is just a generic position denying synthetic consciousness, as used in Blackmore (2004). It is worth noting that Searle himself seemed sympathetic to the possibility of neuromorphic computation being conscious; the target of his original paper is the symbol processing approach to computation, in which computation is the rule-governed manipulation of symbols (see his chapter in Schneider and Velmans, 2017).

5. Searle (1980).

6. See the discussion in Searle (1980), who raises the issue and responds to the reply.

7. Proponents of a view known as *panpsychism* suggest that there is a minuscule amount of consciousness in fundamental particles, but even they think higher-level consciousness involves the complex interaction and integration among various parts of the brain, such as the brainstem and thalamus. I reject panpsychism in any case (Schneider 2018b).

8. For influential depictions of this sort of techno-optimism, see Kurzweil (1999, 2005).

9. This leading explanatory approach in cognitive science has been called the *method of functional decomposition,* because it explains the properties of a system by decomposing it into the causal interaction between constituent parts, which are themselves often explained by the causal interaction between their own subsystems (Block 1995b).

10. In philosophical jargon, such a system would be called a "precise functional isomorph."

11. I'm simplifying things by merely discussing neural replacement in the brain. For instance, perhaps neurons elsewhere in the nervous system, such as the gut, are relevant as well. Or perhaps more than just neurons (e.g., glial cells) are relevant. This kind of thought experiment could be modified to suppose that more than neurons in the brain are replaced.

12. Chalmers (1996).

13. Here I am assuming that biochemical properties could be included. In principle, if they are relevant to cognition, then an abstract characterization of the behavior of such features could be included in a functional characterization.

14. A complete, precise copy could occur in the context of brain uploading, however. Like the case of the isomorph of you, human brain uploading remains far in the future.

CHAPTER THREE: CONSCIOUSNESS ENGINEERING

1. Boly et al. (2017), Koch et al. (2016), Tononi et al. (2016).

2. My search was conducted on February 17, 2018.

3. Davies (2010); Spiegel and Turner (2011); Turner (n.d.).

4. For a gripping tale of one patient's experience, see Lemonick (2017).

5. See McKelvey (2016); Hampson et al. (2018); Song et al. (2018).

6. Sacks (1985).

CHAPTER FOUR: HOW TO CATCH AN AI ZOMBIE

1. Axioms for functional consciousness in highly intelligent AI have been formulated by Bringsjord and Bello (2018). Ned Block (1995a) has discussed a related notion of "access consciousness."

2. Bringsjord and Bello (2018).

3. See Schneider and Turner (2017); Schneider (forthcoming).

4. Of course, this is not to suggest that deaf people can't appreciate music at all.

5. Those familiar with Frank Jackson's Knowledge Argument will recognize that I am borrowing from his famous thought experiment involving Mary, a neuroscientist, who is supposed to know all the "physical facts" about color vision (i.e., facts about the neuroscience of vision) but who has never seen red. Jackson asks: What happens when she sees red for the first time? Does she learn something new—some fact that goes beyond the resources of neuroscience and

physics? Philosophers have debated this case extensively, and some believe the example sucessfully challenges the idea that consciousness is a physical phenomenon (Jackson 1986).

6. Schneider (2016).

7. See Koch et al. (2016); Boly et al. (2017).

8. Zimmer (2010).

9. Tononi and Koch (2014, 2015).

10. Tononi and Koch (2015).

11. I will subsequently refer to this level of Φ, rather vaguely, as "high Φ," because calculations of Φ for the biological brain are currently intractable.

12. See Aaronson (2014a, b).

13. Harremoes et al. (2001).

14. UNESCO/COMEST (2005).

15. Schwitzgebel and Garza (forthcoming).

CHAPTER FIVE: COULD YOU MERGE WITH AI?

1. It should be noted that transhumanism by no means endorses every sort of enhancement. For example, Nick Bostrom rejects positional enhancements (enhancements primarily employed to increase one's social position) yet argues for enhancements that could allow humans to develop ways of exploring "the larger space of possible modes of being" (Bostrom 2005a, p. 11).

2. More and Vita-More (2013); Kurzweil (1999, 2005); Bostrom (2003, 2005b).

3. Bostrom (1998); Kurzweil (1999, 2005); Vinge (1993).

4. Moore (1965).

5. For mainstream anti-enhancement positions on this question, see, e.g., Annas (2000), Fukuyama (2002), and Kass et al. (2003).

6. For my earlier discussion, see Schneider (2009a, b, c). See also Stephen Cave's intriguing book on immortality (2012).

7. Kurzweil (2005, p. 383).

8. There are different versions of the psychological continuity theory. One could, for instance, appeal to (a): the idea that memories are essential to a person. Alternatively, one could adopt (b): one's overall psychological configuration is essential, including one's memories. Herein, I shall work with one version of this latter conception—one that is inspired by cognitive science—although many of the criticisms of this view will apply to (a) and other versions of (b) as well.

9. Kurzweil (2005, p. 383). Brain-based Materialism, as discussed here, is more restrictive than physicalism in the philosophy of mind, for a certain kind of physicalist could hold that you could survive radical changes in your substrate, being

brain-based at one time, and becoming an upload at a later time. For broader discussions of materialist positions in philosophy of mind, see Churchland (1988) and Kim (2005, 2006). Eric Olson has offered an influential materialist position on the identity of the self, arguing that one is not essentially a person at all; one is, instead, a human organism (Olson 1997). One is only a person for part of one's life; for instance, if one is brain-dead, the human animal does not cease to exist, but the person has ceased to exist. We are not essentially persons. I'm not so sure we are human organisms, however. For the brain plays a distinctive role in one's identity, and if the brain were transplanted, one would transfer with the brain. Olson's position rejects this, as the brain is just one organ among many (see his comments in Marshall [2019]).

10. Sociologist James Hughes holds a transhumanist version of the no-self view. See Hughes (2004, 2013). For surveys of these four positions, see Olson (1997, 2017) and Conee and Sider (2005).

11. This is a version of a computational theory of mind that I criticize in Chapter Eight, however. It should also be noted that computational theories of mind can appeal to various computational theories of the format of thought: connectionism, dynamical systems theory (in its computational guise), the symbolic or language of thought approach, or some combination thereof. These differences will not matter for the purposes of our discussion. I've treated these issues extensively elsewhere. (See Schneider 2011).

12. Kurzweil (2005, p. 383).

13. Bostrom (2003).

14. Chapter Eight discusses the transhumanists' computational approach to the mind in more detail.

CHAPTER SIX: GETTING A MINDSCAN

1. Sawyer (2005, pp. 44–45).
2. Sawyer (2005, p. 18).
3. Bostrom (2003).
4. Bostrom (2003, section 5.4).

CHAPTER SEVEN: A UNIVERSE OF SINGULARITIES

1. Here I am indebted to the groundbreaking work by Paul Davies (2010), Steven Dick (2015), Martin Rees (2003), and Seth Shostak (2009), among others.
2. Shostak (2009), Davies (2010), Dick (2013), Schneider (2015).
3. Dick (2013, p. 468).
4. Mandik (2015), Schneider and Mandik (2018).

5. Mandik (2015), Schneider and Mandik (2018).

6. Bostrom (2014).

7. Shostak (2015).

8. Dyson (1960).

9. Schneider, "Alien Minds," in Dick (2015).

10. Consolmagno and Mueller (2014).

11. Bostrom (2014).

12. Bostrom (2014, p. 107).

13. Bostrom (2014, pp. 107–108, 123–125).

14. Bostrom (2014, p. 29).

15. Bostrom (2014, p. 109).

16. Seung (2012).

17. Hawkins and Blakeslee (2004).

18. Schneider (2011).

19. Baars (2008).

20. Clarke (1962).

CHAPTER EIGHT: IS YOUR MIND A SOFTWARE PROGRAM?

1. This quote is from *The Guardian* (2013).

2. Harmon (2015a, p. 1).

3. See Harmon (2015a); Alcor Life Extension Foundation (n.d.).

4. Crippen (2015).

5. I'm grateful to Kim Suozzi's boyfriend, Josh Schisler, for a helpful email and telephone conversations about this (August 26, 2018).

6. Harmon (2015a).

7. Harmon (2015a).

8. Harmon (2015a).

9. Schneider (2014); Schneider and Corabi (2014). For an overview of different steps of uploading, see Harmon (2015b).

10. Schneider (2014); Schneider and Corabi (2014). We never observe physical objects to inhabit more than one location. This is true even for quantum objects, which collapse upon measurement. The supposed multiplicity is only indirectly observed, and it provokes intense debate among physicists and philosophers of physics.

11. For instance, Ned Block (1995b) wrote a canonical paper on this view, titled "The Mind Is the Software of the Brain." Many academics appealing to the Software View are more interested in characterizing how the mind works rather than

venturing claims about brain enhancement or uploading. I will focus on claims by fusion-optimists, as they are the ones making claims about radical enhancement.

12. Wiley (2014).

13. Mazie (2014).

14. Hayworth (2015).

15. Schneider (2011).

16. Descartes (2008).

17. For a helpful new collection on idealism, see Pearce and Goldschmidt (2018). For a discussion of why some versions of panpsychism are forms of idealism, see Schneider (2018a).

18. See, e.g., Heil (2005), Kim (2006).

19. See Schneider (2011b) for a defense.

20. Block (1995b).

21. The notion of an implementation has been problematic for a variety of reasons. For discussion, see Putnam (1967) and Piccinini (2010).

22. Chalmers (1996).

23. Descartes (2008).

24. Putnam (1967).

25. Lowe (1996, 2006). Lowe preferred to speak of the self, rather than the mind, so I am taking the liberty of using his position in the context of a discussion of the mind.

26. Kurzweil (2005, p. 383).

27. Graham (2010).

28. See Schipp (2016).

29. Harmon (2015a).

APPENDIX: TRANSHUMANISM

1. This document appears at the website of the transhumanist organization Humanity+ (Humanity+, n.d.). It also appears in More and Vita-More (2013), an informative volume that includes other classic transhumanist papers. See also Bostrom (2005a) for a history of transhumanist thought.

2. See Bostrom (2003) and Chislenko et al. (n.d.).

REFERENCES

Aaronson, S. 2014a. "Why I Am Not an Integrated Information Theorist (or, The Unconscious Expander)," *Shtetl Optimized* (blog), May, https://www.scottaaronson.com/blog/?p=1799.

———. 2014b. "Giulio Tononi and Me: A Phi-nal Exchange," *Shtetl Optimized* (blog), June, https://www.scottaaronson.com/blog/?p=1823.

Alcor Life Extension Foundation. n.d. "Case Summary: A-2643 Kim Suozzi," https://alcor.org/Library/html/casesummary2643.html.

Annas, G. J. 2000. "The Man on the Moon, Immortality, and Other Millennial Myths: The Prospects and Perils of Human Genetic Engineering," *Emory Law Journal* 49(3): 753–782.

Anthony, A. 2017. "Max Tegmark: 'Machines Taking Control Doesn't Have to Be a Bad Thing'," https://www.theguardian.com/technology/2017/sep/16/ai-will-superintelligent-computers-replace-us-robots-max-tegmark-life-3-0.

Baars, B. 2008. "The Global Workspace Theory of Consciousness," in M. Velmans and S. Schneider, eds., *The Blackwell Companion to Consciousness*. Boston: Wiley-Blackwell.

Bess, M. 2015. *Our Grandchildren Redesigned: Life in the Bioengineered Society of the Near Future*. Boston: Beacon Press.

Blackmore, S. 2004. *Consciousness: An Introduction*. New York: Oxford University Press.

Block, N. 1995a. "On a Confusion about the Function of Consciousness," *Behavioral and Brain Sciences* 18: 227–247.

———. 1995b. "The Mind as the Software of the Brain," in D. Osherson, L. Gleitman, S. Kosslyn, E. Smith, and S. Sternberg, eds., *An Invitation to Cognitive Science*. New York: MIT Press.

Boly, M., M. Massimini, N. Tsuchiya, B. Postle, C. Koch, and G. Tononi. 2017. "Are the Neural Correlates of Consciousness in the Front or in the Back of the Cerebral Cortex? Clinical and Neuroimaging Evidence," *Journal of Neuroscience* 37(40): 9603–9613.

Bostrom, N. 1998. "How Long before Superintelligence?" *International Journal of Futures Studies* 2.

———. 2003. "Transhumanist FAQ: A General Introduction," version 2.1, World Transhumanist Association, https://nickbostrom.com/views/transhumanist.pdf.

———. 2005a. "History of Transhumanist Thought." *Journal of Evolution and Technology* 14(1).

———. 2005b. "In Defence of Posthuman Dignity," *Bioethics* 19(3): 202–214.

———. 2014. *Superintelligence: Paths, Dangers, Strategies.* Oxford: Oxford University Press.

Bringsjord, Selmer, and Paul Bello. 2018. "Toward Axiomatizing Consciousness," http://ndpr.nd.edu/news/60148-actual-consciousness.

Cave, Stephen. 2012. *Immortality: The Quest to Live Forever and How It Drives Civilization.* New York: Crown.

Cellan-Jones, R. 2014. "Stephen Hawking Warns Artificial Intelligence Could End Mankind," https://www.bbc.com/news/technology-30290540.

Chalmers, D. 1996. *The Conscious Mind: In Search of a Final Theory.* Oxford: Oxford University Press.

———. 2002. "Consciousness and Its Place in Nature," in David J. Chalmers, ed., *Philosophy of Mind: Classical and Contemporary Readings.* Oxford: Oxford University Press.

———. 2008. "The Hard Problem of Consciousness," in M. Velmans and S. Schneider, eds., *The Blackwell Companion to Consciousness.* Boston: Wiley-Blackwell.

Churchland, P. 1988. *Matter and Consciousness.* Boston: MIT Press.

Chislenko, Alexander, Max More, Anders Sandberg, Natasha Vita-More, Eliezer Yudkowsky, Arjen Kamphius, and Nick Bostrom. n.d. "Transhumanist FAQ," https://humanityplus.org/philosophy/transhumanist-faq/.

Clarke, A. 1962. *Profiles of the Future: An Inquiry into the Limits of the Possible.* New York: Harper and Row.

Cohen, Jon. 2013. "Memory Implants," *MIT Technology Review,* April 23, https://www.technologyreview.com/s/513681/memory-implants/.

Conee, E., and T. Sider. 2005. *Riddles of Existence: A Guided Tour of Metaphysics.* Oxford: Oxford University Press.

Consolmagno, Guy, and Paul Mueller. 2014. *Would You Baptize an Extraterrestrial? . . . and Other Questions from the Astronomers' In-Box at the Vatican Observatory.* New York: Image.

Crippen, D. 2015. "The Science Surrounding Cryonics," *MIT Technology Review,* October 19, 2015, https://www.technologyreview.com/s/542601/the-science-surrounding-cryonics/.

DARPA. n.d. a. "DARPA and the Brain Initiative," https://www.darpa.mil/program/our-research/darpa-and-the-brain-initiative.

DARPA, n.d. b. "Systems of Neuromorphic Adaptive Plastic Scalable Electronics (SyNAPSE)," https://www.darpa.mil/program/systems-of-neuromorphic-adaptive-plastic-scalable-electronics.

DARPA. 2018. "Breakthroughs Inspire Hope for Treating Intractable Mood Disorders," November 30, https://www.darpa.mil/news-events/2018-11-30.

Davies, Paul. 2010. *The Eerie Silence: Renewing Our Search for Alien Intelligence.* Boston: Houghton Mifflin Harcourt.

Descartes, R. 2008. *Meditations on First Philosophy: With Selections from the Original Objections and Replies,* trans. Michael Moriarty. Oxford: Oxford University Press.

Dick, S. 2013. "Bringing Culture to Cosmos: The Postbiological Universe," in S. Dick and M. Lupisella, eds., *Cosmos and Culture: Cultural Evolution in a Cosmic Context,* Washington, DC: NASA, http://history.nasa.gov/SP-4802.pdf.

———. 2015. *Discovery.* Cambridge: Cambridge University Press.

Dyson, Freeman J. 1960. "Search for Artificial Stellar Sources of Infrared Radiation," *Science* 131(3414): 1667–1668, https://science.sciencemag.org/content/131/3414/1667.

Fukuyama, F. 2002. *Our Posthuman Future: Consequences of the Biotechnology Revolution.* New York: Farrar, Straus and Giroux.

Garreau, J. 2005. *Radical Evolution: The Promise and Peril of Enhancing Our Minds, Our Bodies—and What It Means to Be Human.* New York: Doubleday.

Guardian, The. 2013. "Stephen Hawking: Brain Could Exist Outside Body," *The Guardian,* September 21, https://www.theguardian.com/science/2013/sep/21/stephen-hawking-brain-outside-body.

Giles, M. 2018. "The World's Most Powerful Supercomputer Is Tailor Made for the AI Era," *MIT Technology Review,* June 8, 2018.

Graham, D. W., ed. 2010. *The Texts of Early Greek Philosophy: The Complete Fragments and Selected Testimonies of the Major Presocratics.* Cambridge: Cambridge University Press.

Hampson R. E., D. Song, B. S. Robinson, D. Fetterhoff, A. S. Dakos, et al. 2018. "A Hippocampal Neural Prosthetic for Restoration of Human Memory Function." *Journal of Neural Engineering* 15: 036014.

Harmon, A. 2015a. "A Dying Young Woman's Hope in Cryonics and a Future," *New York Times,* September 12, https://www.nytimes.com/2015/09/13/us/cancer-immortality-cryogenics.html.

———. 2015b. "The Neuroscience of Immortality," *The New York Times,* September 12, https://www.nytimes.com/interactive/2015/09/03/us/13immortality-explainer.html?mtrref=www.nytimes.com&gwh%20=38E76FFD21912ECB72F147666E2ECDA2&gwt%20=pay.

Harremoes, P., D. Gee, M. MacGarvin, A. Stirling, J. Keys, B. Wynne, and S. Guedes Vaz, eds. 2001. *Late Lessons from Early Warnings: The Precautionary Principle 1896–2000,* Environmental Issue Report 22. Copenhagen: European Environment Agency.

Hawkins, J., and S. Blakeslee. 2004. *On Intelligence: How a New Understanding of the Brain Will Lead to the Creation of Truly Intelligent Machines.* New York: Times Books.

Hayworth, Ken. 2015. "Ken Hayworth's Personal Response to *MIT Technology Review* Article," The Brain Preservation Foundation, September 16, http://www.brainpreservation.org/ken-hayworths-personal-response-to-mit-technology-review-article/.

Heil, J. 2005. *From an Ontological Point of View.* Oxford: Oxford University Press.

Hughes, J. 2004. *Citizen Cyborg: Why Democratic Societies Must Respond to the Redesigned Human of the Future.* Cambridge, MA: Westview Press.

———. 2013. "Transhumanism and Personal Identity," in M. More and N. More, eds., *The Transhumanist Reader.* Boston: Wiley.

Humanity+. n.d. "Transhumanist Declaration," https://humanityplus.org/philosophy/transhumanist-declaration/.

Huxley, J. 1957. *New Bottles for New Wine.* London: Chatto & Windus.

Jackson, F. 1986. "What Mary Didn't Know," *Journal of Philosophy* 83(5): 291–295.

Kass, L., E. Blackburn, R. Dresser, D. Foster, F. Fukuyama, et al. 2003. *Beyond Therapy: Biotechnology and the Pursuit of Happiness: A Report of the President's Council on Bioethics.* Washington, DC: Government Printing Office.

Kim, Jaegwon. 2005. *Physicalism, Or Something Near Enough.* Princeton, NJ: Princeton University Press.

———. 2006. *Philosophy of Mind,* 2nd ed., New York: Westview.

Koch, C., M. Massimini, M. Boly, and G. Tononi. 2016. "Neural Correlates of Consciousness: Progress and Problems," *Nature Reviews Neuroscience* 17(5): 307–321.

Kohli, S. 2017. "Bill Gates Joins Elon Musk and Stephen Hawking in Saying Artificial Intelligence Is Scary," https://qz.com/335768/.

Kurzweil, R. 1999. *Age of Spiritual Machines: When Computers Exceed Human Intelligence.* New York: Penguin.

———. 2005. *The Singularity Is Near: When Humans Transcend Biology.* New York: Viking.

Lemonick, Michael. 2017. *The Perpetual Now: A Story of Amnesia, Memory, and Love.* New York: Doubleday Books.

Liao, S. Matthew, ed., 2020. *Ethics of Artificial Intelligence.* New York: Oxford University Press.

Losch, Andreas, ed. 2017. *What Is Life? On Earth and Beyond.* Cambridge: Cambridge University Press.

Lowe, E. J. 1996. *Subjects of Experience.* Cambridge: Cambridge University Press.

———. 2006. "Non-Cartesian Substance Dualism and the Problem of Mental Causation," *Erkenntnis* 65(1): 5–23.

Mandik, Pete. 2015. "Metaphysical Daring as a Posthuman Survival Strategy," *Midwest Studies in Philosophy* 39(1): 144–157.

Marshall, Richard. 2019. "The Philosopher with No Hands," *3AM*, https://www.3ammagazine.com/3am/the-philosopher-with-no-hands/.

Mazie, Steven. 2014. "Don't Want to Die? Just Upload Your Brain," *Big Think*, March 6, https://bigthink.com/praxis/dont-want-to-die-just-upload-your-brain.

McKelvey, Cynthia. 2016. "The Neuroscientist Who's Building a Better Memory for Humans," *Wired*, December 1, https://www.wired.com/2016/12/neuroscientist-whos-building-better-memory-humans/.

MeriTalk. 2017. "DARPA-Funded Deep Brain Stimulator Is Ready for Human Testing," *MeriTalk*, April 10, https://www.meritalk.com/articles/darpa-alik-widge-deep-brain-stimulator-darin-dougherty-emad-eskandar/.

Moore, G. E. 1965. "Cramming More Components onto Integrated Circuits," *Electronics* 38(8).

More, M., and N. Vita-More. 2013. *The Transhumanist Reader: Classical and Contemporary Essays on the Science, Technology, and Philosophy of the Human Future.* Chichester, UK: Wiley-Blackwell.

Müller, Vincent C., and Nick Bostrom. 2014. "Future Progress in Artificial Intelligence: A Survey of Expert Opinion," in Vincent C. Müller, ed., *Fundamental Issues of Artificial Intelligence.* Synthese Library. Berlin: Springer.

Olson, Eric. 1997. *The Human Animal: Personal Identity Without Psychology.* New York: Oxford University Press.

———. 2017. "Personal Identity," in Edward N. Zalta, ed., *The Stanford Encyclopedia of Philosophy*, https://plato.stanford.edu/archives/sum2017/entries/identity-personal/.

Parfit, D. 1984. *Reasons and Persons.* Oxford: Clarendon Press.

Pearce, K., and T. Goldschmidt. 2018. *Idealism: New Essays in Metaphysics.* Oxford: Oxford University Press.

Piccinini, M. 2010. "The Mind as Neural Software? Understanding Functionalism, Computationalism and Computational Functionalism," *Philosophy and Phenomenological Research* 81(2): 269–311.

Putnam, H. 1967. *Psychological Predicates. Art, Philosophy, and Religion.* Pittsburgh: University of Pittsburgh Press.

Rees, M. 2003. *Our Final Hour: A Scientist's Warning: How Terror, Error, and Environmental Disaster Threaten Humankind's Future in This Century—On Earth and Beyond.* New York: Basic Books.

Roco, M. C., and W. S. Bainbridge, eds. 2002. *Converging Technologies for Improved Human Performance: Nanotechnology, Biotechnology, Information Technology and Cognitive Science.* Arlington, VA: National Science Foundation and Department of Commerce.

Sacks, O. 1985. *The Man Who Mistook His Wife for a Hat and Other Clinical Tales.* New York: Summit Books.

Sandberg, A., and N. Bostrom. 2008. "Whole Brain Emulation: A Roadmap." Technical Report 2008–3. Oxford: Future of Humanity Institute, Oxford University.

Sawyer, R. 2005. *Mindscan*. New York: Tor.

Schipp, Debbie. 2016. "Boyfriend's Delivery of Love for the Woman Whose Brain Is Frozen," news.com.au, June 19, https://www.news.com.au/entertainment/tv/current-affairs/boyfriends-delivery-of-love-for-the-woman-whose-brain-is-frozen/news-story/8a4a5b705964d242bdfa5f55fa2df41a.

Schneider, Susan, ed. 2009a. *Science Fiction and Philosophy*. Chichester, UK: Wiley- Blackwell.

———. 2009b. "Mindscan: Transcending and Enhancing the Human Brain," in S. Schneider, ed., *Science Fiction and Philosophy*. Oxford: Blackwell.

———. 2009c. "Cognitive Enhancement and the Nature of Persons," in Art Caplan and Vardit Radvisky, eds., *The University of Pennsylvania Bioethics Reader*. New York: Springer.

———. 2011. *The Language of Thought: A New Philosophical Direction*. Boston: MIT Press.

———. 2014. "The Philosophy of 'Her'," *New York Times*, March 2.

———. 2015. "Alien Minds," In S. J. Dick, ed., *The Impact of Discovering Life beyond Earth*. Cambridge: Cambridge University Press.

———. 2016. "Can a Machine Feel?" TED talk, June 22, Cambridge, MA, http://www.tedxcambridge.com/speaker/susan-schneider/.

———. 2018a. "Idealism, or Something Near Enough," in K. Pearce and T. Goldschmidt, eds., *Idealism: New Essays in Metaphysics*. Oxford: Oxford University Press.

———. 2018b. "Spacetime Emergence, Panpsychism and the Nature of Consciousness," *Scientific American*, August 6.

———. Forthcoming. "How to Catch an AI Zombie: Tests for Machine Consciousness," in M. Liao and D. Chalmers, eds., *AI*. Oxford: Oxford University Press.

Schneider, S., and J. Corabi. 2014. "The Metaphysics of Uploading," in Russell Blackford, ed., *Intelligent Machines, Uploaded Minds*. Boston: Wiley-Blackwell.

Schneider, S., and P. Mandik. 2018. "How Philosophy of Mind Can Shape the Future," in Amy Kind, ed., *Philosophy of Mind in the 20th and 21th Century*, Abingdon-on-Thames, UK: Routledge.

Schneider, S., and E. Turner. 2017. "Is Anyone Home? A Way to Find Out If AI Has Become Self-Aware," *Scientific American*, 19, July.

Schneider, S., and M. Velmans. 2017. *The Blackwell Companion to Consciousness*. Boston: Wiley-Blackwell.

Schwitzgebel, E., and M. Garza. Forthcoming. "Designing AI with Rights, Consciousness, Self-Respect, and Freedom."

Searle, J. 1980. "Minds, Brains and Programs." *Behavioral and Brain Sciences* 3: 417–457.

———. 2016. *The Rediscovery of the Mind.* Oxford: Oxford University Press.

———. 2017. "Biological Naturalism," in S. Schneider and M. Velmans, eds., *The Blackwell Companion to Consciousness.* Boston: Wiley-Blackwell.

Seung, S. 2012. *Connectome: How the Brain's Wiring Makes Us Who We Are.* Boston: Houghton Mifflin Harcourt.

Shostak, S. 2009. *Confessions of an Alien Hunter.* New York: National Geographic.

Solon, Olivia. 2017. "Elon Musk says humans must become cyborgs to stay relevant. Is he right?" *The Guardian*, February 15, https://www.theguardian.com/technology/2017/feb/15/elon-musk-cyborgs-robots-artificial-intelligence-is-he-right.

Song, D., B. S. Robinson, R. E. Hampson, V. Z. Marmarelis, S. A. Deadwyler, and T. W. Berger. 2018. "Sparse Large-Scale Nonlinear Dynamical Modeling of Human Hippocampus for Memory Prostheses," *IEEE Transactions on Neural Systems and Rehabilitation Engineering* 26(2): 272–280.

Spiegel, D., and Edwin L. Turner. 2011. "Bayesian Analysis of the Astrobiological Implications of Life's Early Emergence on Earth," http://www.pnas.org/content/pnas/early/2011/12/21/1111694108.full.pdf.

Tononi, G., and C. Koch. 2014. "From the Phenomenology to the Mechanisms of Consciousness: Integrated Information Theory 3.0," *PLOS Computational Biology.*

———. 2015. "Consciousness: Here, There and Everywhere?" *Philosophical Transactions of the Royal Society of London B: Biological Sciences* 370: 20140167.

Tononi, G., M. Boly, M. Massimini, and C. Koch. 2016. "Integrated Information Theory: From Consciousness to Its Physical Substrate." *Nature Reviews Neuroscience* 17: 450–461.

Turner, E. n.d. "Improbable Life: An Unappealing but Plausible Scenario for Life's Origin on Earth," video of lecture given at Harvard University, https://youtube/Bt6n6Tu1beg.

UNESCO/COMEST. 2005. "The Precautionary Principle," http://unesdoc.unesco.org/images/0013/001395/139578e.pdf.

Vinge, V. 1993. "The Coming Technological Singularity." *Whole Earth Review*, Winter.

Wiley, Keith. 2014. "Response to Susan Schneider's 'The Philosophy of "Her,"'" *H+ Magazine*, March 26, http://hplusmagazine.com/2014/03/26/response-to-susan-schneiders-the-philosophy-of-her/.

Zimmer, Carl. 2010. "Sizing Up Consciousness By Its Bits," *New York Times*, September 20.

INDEX

Page numbers in *italics* indicate illustrations.